A LAND BETWEEN

CENTER BOOKS ON SPACE, PLACE, AND TIME

George F. Thompson, Series Founder and Director

Published in cooperation with the Center for American Places
SANTA FE, NEW MEXICO, AND HARRISONBURG, VIRGINIA

A LAND BETWEEN

OWENS VALLEY, CALIFORNIA

Rebecca Fish Ewan

The Johns Hopkins University Press   Baltimore and London

This book has been brought to publication with the generous assistance of the Graham Foundation for Advanced Studies in the Fine Arts.

The Johns Hopkins University Press
2715 North Charles Street
Baltimore, Maryland 21218-4363
www.press.jhu.edu

Photographs on pages 66, 69 (top), 117 (top), 118, 119 (top), 120–21, 122, 124, 126, 127, 171 (top), 172, 177, 178, and 179 appear courtesy of County of Inyo, Eastern California Museum. Photograph on page 175 is from the Fish family collection. All other photographs were taken by the author between 1991 and 1999.

Library of Congress Cataloging-in-Publication data
will be found at the end of this book.

A catalog record for this book is available from the British Library.

ISBN 0-8018-6460-7   ISBN 0-8018-6461-5 (pbk.)

Frontispiece: Straight road running east to west across the valley.

For my daughter, Isabel Cailin,

    who reminds me every day to go outside and play,

and for my great grandmother Isabel, whose life as a teacher

    and painter of wildflowers inspires me still

CONTENTS

MY FIRST SIGHT of the Owens Valley as I drove up U.S. 395 was magic for me; hit with fingering shafts of light, the mountains, valley, and lake manifested a landscape like no other. Water and light, dust and air washed over the land making it electric, still, and vibrant in the long shadows of late day. Serendipity has guided me throughout my life, and my first excursion to the Owens Valley proved to be my most fortuitous accident. Serendipity runs deep in my genes. My family tree is laden with circuit riders and farmers, people driven by wanderlust and a love of land. My ancestors perpetually traveled westward until they wet their feet on the Pacific shore. They came first from Europe in 1637 and then traveled overland to California in 1862. They, like thousands of others, crossed the Great American Desert looking for gold. They found instead rich soil and a fair climate for growing Lima beans. My great-great-grandfather's bean fields once hugged the coast near Santa Barbara, my birthplace and that of my father, my father's father, and his father's father. It has been written of my great-great-grandfather that "the person who doesn't know Henry

Fish, doesn't know beans." This lineage has given me, a flower child of the sixties long accustomed to shifting with the ebb and flow of life, the ability to stumble upon my own fate while looking for something else.

I came to the Owens Valley under the pretense of completing a vegetation management assignment for a graduate course at Berkeley, picking the valley because I had just read Marc Reisner's *Cadillac Desert* and wanted to see the land that had been so devastated by Los Angeles's water grabbing. I expected to find a place destroyed, a stricken landscape that told the story of L.A.'s thirst and villainy. I found instead not a land marked by one tale of exploit but a region textured with numerous stories of human occupation. This book is a response to the many voices I heard haunting the mountains, fields, and waters of the Owens Valley region.

Perhaps it is my early experiences with the narrow Kern Canyon that heightened my senses to the Owens Valley land and history. The Kern is the only major canyon running north-south in the Sierra Nevada and lies west of the Owens, over the eastern Sierra Range. Lying parallel, divided by the highest peaks in the lower forty-eight states, both valleys are molded from the Sierra's massive blond granite, and both smell of acrid sage, rock, and dust. Formed by faulting and sculpted by glaciers, the Kern Canyon is a deep cleft in the mountains and was once the home of the Tübatulabal people. In 1776, the year San Francisco was founded in Spanish California and independence proclaimed in the new United States on the far eastern reaches of the continent, the river of Tübatulabal country was christened *Rio de San Felipe* by Francisco Garcés, a Franciscan missionary traveling with Juan Bautista de Anza's expedition. The banks of the San Felipe were the farthest east the Spanish ventured in California. In 1845, John C. Frémont renamed the river *Kern* after Edward Kern, an artist and topographer traveling that year on the expedition.

PREFACE AND ACKNOWLEDGMENTS

My great-grandfather, Ben Fish, bought thirty acres beside the Kern River, land that is now wrapped by national forest and park land. It was at the old Fish Camp that I first fell in love with rivers, land, and history. On my last trip to the Kern while working on this book, I sat with my feet dangling in the cold, high mountain waters of the river in this thin canyon and heard the voice of this book played out in the waters before me. Across the river a small stream cascaded down the steep bank. Hidden mostly among sneezeweed, groundsel, paintbrush, and cottonwood saplings, the water turned silver in a shaft of sunlight that broke through the ponderosa pine and incense cedar that line the banks and disappeared behind a barkless tree trunk dropped beside the river when high spring waters receded. The stream divided into mirrored drips feeding into the river, small, quiet melodic drops all in a line running down the water-bleached log. This stream, with hundreds of others, gives its waters to the Kern, yielding its singular notes to the river's music. Making this book has been like trying to put to paper a landscape's score, an exercise more in listening than in writing. Each person who dwelt in or visited the valley has a different story to tell, and I have tried to listen carefully to the many voices.

When asked what the book is about I usually falter, because I haven't yet found a way to say land, water, history, nature, culture, love, and loss in one breath. It's as much a natural history as one of culture, a book about the intricate ties between people and place. During the writing I've often envied natural history writers and cultural geographers, because the former are focused on nature and the latter are focused on humans (there are many hybrids of the two disciplines, but I feel more kindred than envious of these blended folks). This book fits between these disciplines; being so wedged in the space between nature and culture, really more of an ecotone than a distinct divide, the writing of it not

xiii

only inspired the title but also helped me find my own field of beans.

The book's introduction describes the physical setting of Owens Valley and examines first impressions of the land, and it considers how the circumstances of travel effect these initial perceptions of the valley. The first essay, "A Land Between," explores the valley's natural history and begins to make the connection between the ecology of place and human use. The second essay, "Dwelling Before," chronicles the major periods of human occupation, beginning with the Numu (also refered to as Owens Valley Paiute in many sources) and ending when the Los Angeles Department of Water and Power first spilled the valley's river water into the Los Angeles aqueduct. The third essay, "Lives Diverted," considers life in the valley after diversion.

The words have emerged from the land, from places to which I have been. The heart of this work was written on horseback alongside meadows on the Kern Plateau, while hiking along the Owens Valley bottomlands and the Sierra's granite alpine peaks, on old stones scattered in the Inyo foothills and the Alabama Hills, on the summit of Mount Whitney, perched on narrow divides between the steep eastern drop to the valley and the ancient meadowlands of the Sierra, in hotel rooms, while driving along U.S. 395, beside remnant gardens of Manzanar's ruins, along empty dirt roads running past the Owens River, beside graves and workings of miners at Cerro Gordo, among the 4,700-year-old bristlecone pines in the White Mountains, and sitting on rocks beside the Kern River. It was written in the land and is best read there. It is my hope that readers will carry this book with them to the Owens Valley, sit on a rock in the Alabama Hills, or hike to Mount Whitney or elsewhere in the Sierra or the White-Inyo Range to find a quiet place to read. The words attempt to bring out the sounds, smells, and history of this sacred land—a land of water and light, a land of layered human history. The book is meant to be an appendix to the landscape itself.

My geographical focus is on the southern portion of the Owens Valley in eastern California, though it drifts into the Sierra Nevada and White-Inyo mountain ranges, as well as northward to Bishop and the Round and Long Valleys. The mountains not only give the valley structure, they also influence the natural and cultural history, making them integral to dwelling in the valley. Like a narrow version of the Western frontier, the region has a land-use history that illustrates landscape changes in the arid American West. Although the valley has undergone numerous changes since 1800, three distinct periods of landscape and human history are notable, because they are separated by extreme changes in both land use and landscape perception. The first period started well before the nineteenth century and represents occupation by the Numu people. They lived undisturbed in the valley, except for occasional skirmishes over territory with neighboring native peoples, until the middle of the nineteenth century. Conflicts with stockmen over land mounted, peaking in 1865 when the United States military and settlers suppressed the Numu. After 1865, a few minor fights occurred, but ranchers and farmers now managed the land. This agrarian period lasted until 1913, when the city of Los Angeles diverted the Owens River water to the San Fernando Valley via the Los Angeles aqueduct. Land use has been eclectic since that time.

I relied on primary sources as often as possible so that the landscape history could be described in the words of valley dwellers and visitors. These sources help reveal the conflicts in landscape perception that are at the heart of the conflicts between different groups of people. The valley has at times been seen as a verdant oasis when in truth it is dominantly desert scrubland receiving less than ten inches of rain per year. Land-use decisions appear to have been based more on landscape perception than on measurable conditions of the valley. Differing landscape perceptions also gave rise to social clashes between communities, for example between the Numu and ranchers, in which the former group's homeland was seen by the latter as an untamed frontier. For the nineteenth-century historical information, I have been limited to perceptions of those who wrote about the valley. Abundant literature for the agrarian and postdiversion periods exists, but I found little on how the Numu perceived the valley prior to European American and Mexican contact.

Throughout the book I have integrated the text with personal and descriptive writing to balance more objective observation with my own visceral reaction to the valley landscape and history. The sans serif text is either from my field notes or my fictional writing, done in response to studying landscape experiences of the past. The fictional pieces in "Dwelling Before" I wrote after reading about the Numu, miners, and nineteenth-century settlers and imagining what it may have been like to live in the Owens Valley during periods of their dominant occupation. Between each essay, the galleries reveal the Owens Valley as a land of contrasts, of layered and diverse human history, and of change.

When I was an undergraduate at Berkeley, Professor Gunther Barth told his students, myself among them, that learning history is only valuable if it effects change today, if the lessons learned are of use in improving life in this moment. His words, no doubt repeated every semester to a new batch of Cal juniors, gave history credibility to me. I understood the point of remembering the past; it is to help create a better *today*, not a richer tomorrow or a more palatable yesterday. I have kept his words close to my heart while writing this book. By wandering through histories, reflections, and yellowed newspapers, I've glimpsed the fleeting evolution of the West as it played out in the Owens Valley and tried to create a narrative that casts light on the valley as it is today, to reveal it as a landscape threaded deep with lore and stories that affect how the valley is lived in and seen today. I've grown up Western yet never fully grasped the significance of this fact until I began this work; I took for granted the length of my roots in California and the stability they give me. (Growing up along the San Andreas Fault, I should have known better.)

The landscape history of the Owens Valley reveals that stability in the West is both precious and fragile; the relationship between people and the land is deep and often passionate, yet the balance of this union can be shaken overnight. This narrow valley between the White-Inyo Mountains and the Sierra, a place that at first glance might seem like just another sleepy rural community, has been perpetually jostled. The epilogue, "Parting Glances," reflects on the changes in the land and how history continues to evolve in the valley. In the epilogue I also revisit how changes in myself have affected how I read and understand the land and its history, making clear the connection between story and storyteller. Having dialogue with the land and telling its story are gifts that come naturally to young children, gifts that too often are lost as children grow up. The joy I experienced in writing this book came from being able to nurture this gift in myself and to share it with others. It is my hope that the book may help others re-engage with the land.

I WOULD LIKE to thank the director, Bill Michael, and museum specialist, Beth Porter, at the Eastern California Museum, for helping me locate oral histories of Owens Valley residents and photographs of the valley. Arthur Hansen, director of the oral history program at Cal State Fullerton, also helped me find the voices of those who had been interned at the Manzanar War Relocation Center. Kari Coughlin of the National Park Service gave me a good understanding of how people perceive the Manzanar Historic Site. I'm grateful to Freida Brown and Ellen and Erin Howard at the Paiute Shoshone Indian Cultural Center, for taking the time to review my list of Numu words. Chris Plakos and Linda Ellis, at the Bishop office of the Los Angeles Department of Water and Power, and Theodore Schade, of the Great Basin Unified Air Pollution Control District, helped me sort out the current relationship between L.A., the valley, dust, and water.

I acknowledge the kind hospitality of the people of the Owens Valley, especially the owners and employees at the local cafes and motels. Jody Stewart and Mike Patterson helped bring mining in the valley to life during my stay at Cerro Gordo, my first field excursion with my toddling daughter, Isabel. Many thanks as well to the Rock Creek Pack Station, especially owners Craig and Herbert London, and their knowledgeable packer Jim Brumfield, a man almost as hard-headed as myself, who led me through the southern Sierra and into the history of grazing on the Kern Plateau. Along the trail, Cindy Wood, a National Park Service ranger, and Jim Harvey, the head packer for Sequoia National Park, helped me understand the history of the region. I am also grateful to the cattlemen of the Kern Plateau who leaned over split-rail fences to talk to me about cows.

I would like to thank the many teachers in my life who have encouraged me and are an undying source of inspiration, in particular, Kathleen Aldrich, Stanley Cardinet, Gretchen Griswold, James Kelly, David Kinstle, Susan Matthews, Charles Souza, Clare Cooper Marcus, Chip Sullivan, and Joe McBride. I also would like to thank my father, Peter Morton Fish, uncle Henry McLauren Brown, and great-grandfather Benjamin Harrison Fish for introducing me to the southern Sierra. They instilled in me as a young child a deep love of the mountains and a strong land ethic, and they taught me about glacial moraines, packing a mule, avoiding rattlesnakes, and gutting a trout. While working on this book I re-united with my uncle after years without visits for countless reasons, none very reasonable, and was able to listen again to stories of the mountains. *Vaya con Dios*, Henry.

Endless thanks to George F. Thompson, president of the Center for American Places, for his continual support and thoughtful editorial advice. Celestia N. Ward, manuscript editor at the Johns Hopkins University Press, was sensitive and thorough in her review of the manuscript. I have benefited, also, from others who have read the text, among them Joe Ewan, Randall Jones, Cotton Mather, Paul Starrs, and Frederick R. Steiner. Allyce Hargrove was a tremendous help to me with the maps. The School of Planning and Landscape Architecture at Arizona State University was supportive, particularly with photographic processing and field work. My deepest gratitude to the Graham Foundation for Advanced Studies in the Fine Arts for its support of my field work; the book is more complete because of the foundation's generosity. And always, for everything, my husband Joe . . . thanks.

xix

The greater portion of this immense region, including Owen's Valley, Death Valley, and the Sink of the Mohave, the area of which is nearly one fifth that of the entire State, is usually regarded as a desert, not because of any lack in the soil, but for want of rain, and rivers available for irrigation. Very little of it, however, is desert in the eyes of a bee.

JOHN MUIR, *THE MOUNTAINS OF CALIFORNIA*, 1894

A LAND BETWEEN

INTRODUCTION FIRST IMPRESSIONS

The perceptions of any people wash over the land like a flood,
leaving ideas hung up in the brush, like pieces of damp paper to be
collected and deciphered. No one can tell the whole story.

BARRY LOPEZ, "THE COUNTRY OF THE MIND"

HOLISTIC LANDSCAPE interpretations speak in both general and specific voices and are as universal as the physical laws that help to explain a mountain, as unique as a single human being living at its rocky feet. Interpreting landscape history requires one to listen to all this speaking, which is muffled and layered in time. What follows is one such interpretation for the alluring landscape called Owens Valley, a place rich in natural and cultural history, steeped in story as deep and varied as its alluvial foundation. This brief exploration visits places in the southern end of the valley, including Owens Lake and the towns of Independence, Lone Pine, and Keeler. It stretches north to Bishop, the Round and Long Valleys, up and over the mountainous western edge to the Kern Plateau, to Mount Whitney peak, east to Cerro Gordo and the White-Inyo Range. It meanders the streams that feed the Owens River and lingers by the river. It considers the territorial lands of Numu villages such as Pitana patu, Panatu, and Tovowahamatu. Owens Valley sits east of the Sierra Nevada, a 210-mile drive northeast of Los Angeles, and 231 miles west of Las Vegas. It neighbors Death Valley, the lowest point in the United States, which lies just 108 miles east on State Route 190 over the Panamint Range. This narrative relies on stories told by the land, remnants of human and natural history, books, old photographs, films, oral histories, and conversations. These stories are threaded together not so much to communicate one linear history as to convey the tapestry made by centuries of human occupation and wandering in the region. As Lopez says, "no one can tell the whole story." There is in fact no whole story, no concise tale tucked neatly between "once upon a time" and "they all lived happily ever after."

Geologic events prepared for the shaping of the Owens Valley as far back as 600 million years ago, when oozy benthic layers formed beneath an ancient sea, but the valley's landscape genesis, when people began to shape the land, occurred with human emigration, at least 8,000 years ago. Human movement during this period is evidenced by weapon points found in the valley, left by people who may have been the ancestors of the Numu, the native people who occupied the valley in the nineteenth century, when written records of it began.

This story of landscape thus begins as remembered in Numu myths. The first of the following origin myths was recounted by Hoavadunuki, a man who lived in Tovowahamatu (meaning *natural mound place*), a village on the site of present-day Big Pine. It reveals a less arid environment and accurately notes remnants of human occupation in the high mountains:

> The world was once nothing but water. The only land above the water was Black mountain. All the people lived up there when the flood came, and their fireplaces can still be seen.

Fish-eater and Hawk lived there. Fish-eater was Hawk's uncle. One day they were singing and shaking a rattle. As they sang, Hawk shook this rattle and dirt began to fall out of it. They sang all night, shaking the rattle the whole time. Soon there was so much dirt on the water that the water started to go down. When it had gone all the way down, they put up the Sierra Nevada to hold the ocean back. Soon they saw a river running down through the valley.

When they finished making the earth, Hawk said, "Well, we have finished. Here is a rabbit for me. I will live on rabbits in my lifetime." Fish-eater was over a swampy place, and he said, "I will live on fish in my lifetime." They had plenty to eat for themselves. It was finished.[1]

The second myth describes the consequent drying of the land that, according to legend, was precipitated by angry bears. It was published in 1909 in *Sierra Magazine* (formally the *Inyo Magazine*) and credited to a man referred to in the article as American Joe, who had told it in 1867. Its mythic quality is no doubt culturally layered by nineteenth-century journalistic license:

The bears became very angry. They burrowed a great hole under the lake, and filled it with wood from the forest, and set fire to the wood—and a great fire arose from the valley. It dried up the lake, and the snows from the mountains, and burned off the forest, so that all the country became bare. The lake gone, there were no more fish nor wild fowl for the Indians to live on, and they went away. After a long time, the Great Spirit planted pinyons on the hills, and the Indians came back; for now they had learned to live on pine nuts alone.[2]

According to these myths, human occupation in the valley probably began in the late Pleistocene epoch. By this time, the eastern range of the Sierra Nevada had formed the valley's western wall, the White-Inyo Mountains had upthrust to the east, and the valley floor had dropped below sea level and filled with alluvium thousands of feet deep. Volcanism had long since calmed, leaving red cinder cones and rough-hewn black lava flows scattered about the valley landscape. A waning ice age had carved sharp peaks from the Sierra Range, its receding glaciers leaving piled rubble moraines, high azure lakes, and deep rounded valleys in the mountainous terrain. And down the steep Sierra slopes, snowmelt waters flowed to form in confluence the Owens River that fed the then vast Owens Lake before spilling into neighboring valleys. About 6,000 years ago, water stopped overflowing the volcanic ridge that is the southern terminus of the valley. The lake level may have risen high enough during the Little Ice Age (mid A.D. 1500 to mid A.D. 1800) to once again breach the lava flow. Except for this brief cooling period, the Earth's climate has been warming since the Pleistocene, and in the valley, even without human interference, the lake water has been evaporating faster than it has been replenished. So the valley has the structure and geomorphology as it is described in the first myth and experienced the climatic shift noted in the second. So it is today. So it was when nineteenth-century visitors wrote down their first impressions and experiences while passing through the valley on their way to somewhere else.

Early descriptions recorded in journals and often augmented with recollections before publication were made by men traveling through the valley, none of whom had any intention of staying. They, like myself, were visitors, and their journals and letters read in this context are better understood. Those who dwelt and continue to dwell in the valley, whose words dot the three essays that follow, are profoundly rooted to the land and speak from a depth that makes visitors want to linger, in the hope that they might begin to feel the land grip them like it has the dweller.

In 1834, Zenas Leonard, a trapper from Clearfield County, Pennsylvania, made one of the earliest recorded trips through the valley while employed as a clerk on a fur-trapping expedition led by Captain Joseph Reddeford Walker. Like most travelers in the West of his time, Leonard covered great distances over many months with little comfort and relied on horses and other stock animals for transportation and food; his impression of the valley reflects both his weariness from ten months of travel and his constant concern for good pasturage for the animals. The group entered the southern end of Owens Valley in early May, "reduced to 52 men, 315 horses—and for provisions, 47 beef, and 30 dogs," after leaving the beautiful San Joaquin Valley and toiling for four snowy days to cross the Sierra Nevada. After spending time in the well-watered San Joaquin Valley, where the "prairies were most beautifully decorated with flowers and vegetation, interspersed with splendid groves of timber along the banks of the rivers—giving a most romantic appearance to the whole face of nature," his disappointment with the Owens Valley's dry landscape is palpable. He measured the two river valleys against each another, and the Owens came out the clear loser. "The country on this side is much inferior to that on the opposite side—the soil being thin and rather sandy, producing but little grass, which was very discouraging to our stock, as they now stood in great want of strong feed."[3]

The group was heading back to the United States, which at the time had its western border in Missouri, so the men were tired and, like horses who have turned toward their corral, eager to make progress. The trek north up the long, narrow valley was a burden to be endured since the States lay eastward. "Traveling along the mountain foot, crossing one stream after another, was anything but pleasant. Day after day we travelled in the hope each day of arriving at the desired point when we would strike off in a homeward direction. Every now and then some of the company would see a high peak or promontory, which he would think was seen by the company on a former occasion, but when we would draw near to it our pleasing anticipations would be turned into despondency."[4] After leaving the valley, they were so anxious to be home that they made a premature and failed attempt to cross the Forty Mile Desert to reach the Humboldt River, during which they lost sixty-four horses, ten cows, and half their dogs to heat and dehydration. Thirst drove the men to drink the blood of the dead animals. Perhaps the valley would have seemed more promising to Leonard had he traveled the other way, had he come upon it after toiling to cross the desert rather than meeting it on his return home.

Joseph Reddeford Walker led the Chiles party of emigrants down the valley in 1843, but no record was made. The diarists in the party split off before reaching the valley, one traveling on to Oregon because he didn't think the Chiles party had sufficient provisions and the other detouring on horseback with a group of the party's strongest men to try to round up food to augment the low supply. Walker had kept a journal during his years of travels, but it was unfortunately lost in the current during a river crossing. It is known that the party that entered the Owens Valley was reduced to twenty-five, including five children, and only one person besides Walker had much experience traveling. They had waited at the rendezvous point beside the Humboldt Sink, the last decent pasturage before the Forty Mile Desert crossing, without a sign of the young men for as long as Walker felt their dwindling provisions could bear. They then moved south along the Sierra Range. Abandonment of their wagons, the first wagons to make the overland trip into California, on the shore of Owens Lake in late November speaks for their condition at that point in their travels. They cached their belongings, including a full set of mill irons, a weighty and valuable bit of cargo that even John C. Frémont noted in his memoirs, burned the wagon wood for warmth, and continued on safely over Walker Pass (elevation 5403 feet).

In December 1845, Theodore Talbot and Edward Kern both traveled, again under the guidance of Joseph Reddeford Walker, through Owens Valley. The military exploration on which Talbot and Kern journeyed through the valley, Frémont's third such expedition, had begun in June in present-day Kansas City, Missouri. The company divided at the Humboldt, and Talbot was put in charge of the group that went south along the base of the Sierra and over Walker Pass to the San Joaquin Valley. By the time Talbot's company reached the Owens Valley, rations were low and most of the men were walking so as not to further tire their weary animals. Talbot recorded the journey in letters home to his sister and mother, and Kern kept a journal.

Talbot, a Catholic from Kentucky, was well educated and enjoyed the company of distinguished men and good food. His only comment on the valley and surrounding landscape was dreary: "Most of the men were afoot our food horse meat and the country sublimely desolate affording no game and inhabited by hostile indians. We spent a rather gloomy Christmas and on the 27th of December reached our appointed rendezvous."[5] Kern, an artist from Philadelphia and the last of nine children, having just turned twenty-two, was anxious to cross the Sierra to more plentiful land and made a lukewarm commentary on the valley. His first entry along the Owens River on December 16 notes the river's "fine, bold stream," but he comments more on the company's hunger and desire to be elsewhere. "Our rations are becoming extremely scant. The men being all on foot, they feel their appetites much quickened by the additional exercise of walking. A few more days we hope will bring us to the land of plenty."[6] As they were traveling with Walker, a seasoned guide in the West who knew more about this region than any other European American of his time, they would be well aware of what lay ahead in the San Joaquin Valley and other regions of California, places where game was plentiful and where they would no longer have to make their meals from horse and mule.

These few early accounts saw little beauty or value, at least none worth their record, in the Owens Valley landscape. Nothing of remark, save the continuation of their weary travails, remains in their writing. The valley's closeness to verdant land just over the mountains that was thick with game and rich pastures seemed to make the landscape's least appealing qualities more tangible for these men, and their Christmas dinner ("by way of a change, on one of [their] tired, worn mules, instead of a horse") seem all the more dreary. While Kern blandly recorded the land east of the Sierra—"Grass poor," "Ducks and geese plentiful"—his account of the group's first sight of the San Joaquin Valley is one of his most exuberant. On the summit he had "the first view of the much-wished-for Valley of California. It lay beneath us, bright in the sunshine, gay and green, while about us everything was clothed in the chilly garb of winter." The descent took them through "beautiful groves of live and other oaks, clear from growth of underwood; the fine grass [giving] the country the appearance of a well-kept park."[7]

When gold was discovered at Sutter's Mill in 1848, the American emigrant population that had up to then been a trickle became a torrent. California joined the Union in 1850 after the war with Mexico, further facilitating the flood. The Owens Valley shifted its role from part of the Great American Desert to a piece of the new frontier, albeit a remote one. The government began to send people into the valley to survey the land and investigate the native population. Perhaps because people were looking for potential in the land rather than seeing the valley as an obstacle, accounts of subsequent ventures into the valley are more positive, with the exception of A. W. Von Schmidt, who surveyed the valley in 1855–56 and considered the land worthless.

Captain John W. Davidson made an expedition from Fort Tejon in 1859 to investigate allegations that the Numu were stealing horses from the area around Mission San Fernando. He was also ordered to study the land to evaluate "its fitness for the purposes of an Indian Reservation, its Agricultural resources, timber, water, &cc."[8] From Fort Tejon—a military base in the Tehachapi Mountains established in 1854 to oversee the San Sebastian Reservation and protect travelers on the road between Los Angeles and the San Joaquin Valley—Davidson took only nine days to cover the 175 miles to Owens Lake. Compared to the other travelers through the valley, his expedition was brief and his accounts are not burdened with fatigue. Even at the southern end of the valley, where the land is more arid, he noted the "fine mountain streams" and "luxuriant meadow." Traveling up the valley, he remarks that "every step now taken shows you that nature has been lavish of her stores. The Mountains are filled with timber, the vallies with water, and meadows of luxuriant grass." Just north of present-day Independence near Oak Creek, he noted the "grove of magnificent oaks" and from his perspective looking east to Owens River "the eye wanders over a sea of green."[9] Clearly his perception of the valley differs from those of the earlier travelers. Moreover, his is the last recorded exploration before the valley becomes populated with emigrant farmers, ranchers, and miners. Philip J. Wilke and Harry W. Lawton, the editors of his report, speculate that his buoyant descriptions of the landscape may have been a catalyst for the flood of settlers. Perhaps, but it may also simply have been that

11

California's overland emigrant population went from a handful in 1847 to over 165,000 ten years later, with many more coming by sea to San Francisco. People were beginning to seek out land and mining prospects in the more remote reaches of the state.

One hundred and forty-two years after Davidson's expedition, intrigued by accounts I'd heard of the Owens Valley and the water wars that were still lingering, I made my first visit to the valley. Like these men before me, I came as a visitor, and my impressions, and indeed this book, are colored by the temporariness of my experiences.

MARCH 24. First visions. Driving up U.S. 395, listening to Randy Travis sing his promise to find a place for me somewhere in his broken heart, I watch the approach to Owens Valley by the rising white Sierra Nevada to the left and the encroaching brown Inyo Mountains on the right. Joshua trees and creosote thin and shrink away in deference to blackbrush and saltbush. A strong wind blowing up the valley, making alkali dust cloud thick over the lake, pushes my truck along. Desert vegetation, adapted from millennia of fierce wind, gives little evidence to the constant gales. Small mounded shrubs stand peculiarly still almost in defiance to the wind, denying the powerful gusts and turning attention to a more urgent concern—the search for water. Aridity shapes the vegetation, spreading it thin and taut across the dry land, leaving moist wrinkles along streams that pour down each western canyon and meander like snakes across the vast bajada to the valley below. Roots stretch beneath the dry soil, like thirsty drunks at last call. And wind drifting over the looming Sierra Nevada sheds the last remnant moisture from the air as it pushes east. When rain falls, it comes with infrequent yet violent bursts upon the bottomlands.

Driving past Owens Lake I marvel that the cows here have learned to walk on water. A closer look reveals a lake turned to dry fodder and dust. Cows, ignorant of the change, perhaps even grateful, stare blankly chewing and crapping amongst the saltbush, oblivious of the dying lake and their own imminent slaughter. Yawn, fart, chew, and move on, a dull example of American living.

My first vision of Owens Valley is molded by those who came before, in the days when cows, miners, native peoples, farmers, and Los Angeles engineers scraped across the valley floor, each leaving new traces while erasing old ones and all finding their salvation in the waters of Owens River. I yearn to breath lore out of the dry air, to carry it back in quick sketches; to lay my expectations on the landscape as others have before me, hoping to make a big strike and carry home a profound nugget from this remote valley. Am I panning for fool's gold, missing genuine truth, because my eyes cloud with preconceptions and overblown romanticism about the Old West? Will dusty valley characters named Shotgun Charlie ramble across my path, because I was weaned on the Lone Ranger and taught to love men who smell of leather, sweat, and whiskey?

Later, nestled comfortably into a room at the Winnedumah Inn on Highway 395, I listen to the intermittent rhythm of cars driving through town, heading north to Mammoth or south to Los Angeles. The inn is a rattley old place owned for years by a woman named Hattie Schaefer, who, according to the clerk at the Minimart, was a fine woman, like everybody's grandmother, full of stories and laughter. The building groans. When she died, her inn sagged with mournful regret. Wind whistles in the vents, banging the old wooden window frames; I shiver from the thought of the cold outside and tuck deeper into the musty woolen blankets. The night air breaths crisp and full of stars. As a newcomer, I can't read the signs of the coming storm predicted by locals in the cafe across the street—I only note the pleasant, calm morning breeze followed by powerful, unrelenting afternoon winds. Gusts blow up the valley over the drying remains of Owens Lake, sending white alkali dust into a big haze over the flat terrain. We're promised new snow in the morning.

MARCH 25. Kearsarge Station. Mary Austin's words echo in the biting wind. The old picket fence running nowhere, dividing nothing, is reminiscent of the neighbor's fence she wrote of in *Land of Little Rain.* But Mary Austin's Kearsarge sits elsewhere, beside the bubbling Pine Creek, cooled in tree shade and the mingled sound of water and children's laughter. Not here, where the dry McIver Canal runs past Mazourka Canyon Road and behind scant traces of buildings. A USGS map shows three structures in Kearsarge Station, but only remnant foundations and cellars lay scattered among sand and desert scrub. A bullet-ridden stump suggests a once-shaded yard beside the canal, but Kearsarge Station is now a place of shallow graves for dogs and old refrigerators.

I sit in my truck to escape the bitter cold wind. Two black crows fly low above the brush and dip out of sight. Saltbush trembles in the wind, holding fast in the sandy soil. Beyond lie the Inyo Mountains, with brown hummocky foothills pushing out onto the valley floor, their faces steep and scarred. A road cuts clear up the nearest face, its beginning and end blurred in folding topography. Like the picket fence, the road stands in anachronistic illogic, seeming to lead nowhere from nowhere, its purpose lost in time so only form remains.

Every marking lasts in this dry land. The entire landscape bears scars and wrinkles from endless seasons of human activity and the molten unrest that left volcanic bumps and folds all along the long thin valley. Everything lasts except water. It blows away with the wind, leaving dry sandy earth sparsely dressed in the scratchy garments of sage and saltbush. Hues of black shadows, ruddy coppery streaks, milky greens, and steel grays brush over eroded slopes. Despite the rugged desert scrub, vegetation made soft by distance appears as a velvet cloth draped over even the nearest mounded hill.

Beyond the Inyo foothills, mountains stretch into forest elevations, changing from dry dusty brown to deep inky green and snowy white. Storm clouds hang behind the mountains as if waiting to ambush the thirsty valley soil.

To the north, rain mist sweeps across the valley casting blue-gray shadows on volcanic soil below. The Sierra Range loses its peaks in these stormy vapors, but here in the Kearsarge Station ruins, rainless wind dances upon the still ground. Occasional heavy gusts carry sand over from Owens Lake to bury the dog, Duke, deeper in his grave and blot out the rusty broken vestiges of Kearsarge. The station is as dead as Duke and no doubt less sorely missed. Once a stop for the Carson & Colorado narrow gauge trains, Kearsarge is one of many rotting traces of the mining boom that swept through the desert regions in the late 1800s.

Yet it is a lovely place in a sad and desolate way, speaking with a lonely dry voice, talking of death and mourning. The wind murmurs and the land recalls Duke, the Numu, old hairy shepherds that camped here, and the cottonwoods that once cast happy shade in the station yard. I love the lonely isolation, the rusty remains of olden times. These scattered ruins hint of bustling yesteryears, making the land feel empty and expansive.

Although first impressions such as these tend to linger and influence further perceptions, Owens Valley history shifts like the cloud shadows across the Inyo. With each returning, each idle comment made by a clerk who sells me chocolate, the narrative changes. There is no universal history, no one story, only perceptions, memories, and markings on the land. So reading the valley landscape is like turning the pages of a book that rewrites itself over time, coming back again never reveals the same tale. This fact of landscape history makes searching for the beginning and the end futile. Like the fence at Kearsarge Station, the story begins and then ends in a different place after every new storm shifts the sand.

History in general is a reinventing of the past to give sense to the present. This is especially true of landscape history when read in the dynamic folds of the earth. Since landscape embodies betweeness—the place between heaven and earth, between time, between cultures, between perception and reality—it is a construct wedged between disciplines and cannot be seen as either a purely historical product or one of purely contemporary design. Owens Valley is a land between, a place tucked behind high mountains, arid yet soaked in water history, draped in desert vegetation yet remembered for its verdant farms, sparsely dotted with towns— some no more than dreams on a map. It exists between stories, between vitality and decline, between granite mountains.

16

1   A LAND BETWEEN

THE OWENS VALLEY is flanked by two magnificent mountain ranges—the Sierra Nevada and the White-Inyo. Separated by the narrow valley, proximity magnifies their contrasting character. Most written accounts describe the strong presence of the Sierra Nevada rising more than 10,000 feet above the valley, snow-crested and austere, craggy peaks that emanate power. The Sierra Nevada began as a series of magmatic intrusions into the earth's upper crust as the Farallon plate, which lies beneath the Pacific Ocean, squeezed under and melted below the North American plate. Granite batholiths, immense formations of stone, developed and, joining with older intrusions, created an expansive rock floor. The stone that cradles the valley is of the older batholiths, formed in the Jurassic Period between 170 and 200 million years ago. The Owens Valley itself formed through a series of uplifts and downthrows beginning over 25 million years ago as the land tilted westward, leaving the steep faulted relief of the Sierra's eastern face. Glaciation during the Pleistocene Ice Age (2 million to 11,000 years ago) sculpted the massive granite blocks into sharp peaks, rounded valleys, and bowled and often lake-filled cirques. The resultant mountains of the Sierra's eastern flank are sawtoothed and snow-capped granite monuments. They seem still and eternal, yet they continue to flux and fault. The uplift is the fastest it has been in 25 million years, raising the mountains quicker than weathering can erode them.

People often feel the presence of a Higher Being in the mountains. John Muir, who spent much of his life rapt in the beauty of the High Sierra, saw the alpine light as translucent divinity draping itself across the mountain peaks each day at dusk. "Long, blue, spiky shadows crept out across the snow-fields, while a rosy glow, at first scarce discernible, gradually deepened and suffused every mountaintop, flushing the glaciers and the harsh crags above them. This was the alpenglow, to me one of the most impressive of all the terrestrial manifestations of God. At the touch of this divine light, the mountains seemed to kindle to a rapt, religious consciousness, and stood hushed and waiting like devout worshipers."[1] The mountains stand in a jagged row—including the fourteeners, Langley, Muir, Whitney, Russell, Williamson, Tyndall, Split, Middle Palisade, and Sill, whose peaks reach over 14,000 feet—each one a giant lost among giants. Their scale humbles the human activity at their feet, yet, since all is largeness, the mountains dwarf one another, making them seem simultaneously monumental and approachable. Their immensity often challenges people to attempt to conquer and tame what is and always will be larger than themselves.

Climbing to the summit of Mount Whitney (elevation 14,496 feet), the highest peak in the lower forty-eight states, embodies this pursuit of monumental conquest. The ascent has evolved from the dramatic climb described by Clarence King, a nineteenth-century geologist who had a singular, and ultimately unsatisfied, passion to be the first to reach the Whitney summit, to the popular hike now made by thousands each summer. King's ascents of mountains he mistook for Whitney were more treacherous and sublime, but even his successful climb in 1873—among the third party to reach the summit—is described with enough spice to appeal to the nineteenth-century taste for the horrible. He and his companion were faced with snow fields that "often gave way, threatening to hurl [them] down into cavernous hollows." At one point he falls and must creep "on hands and knees up over steep and treacherous ice-crests." In the end, recollecting the climb, he admits that it was not so dangerous, though he goes on to qualify this admission, noting that the danger was less for those with sufficient mental fortitude and physical strength. "The utter unreliableness of that honeycomb and cavernous cliff was rather uncomfortable, and might, at any moment, give the death-fall to one who had not the coolness and muscular power at instant command."[2]

King noted the westward view to the Kern Canyon more than that east to Owens Valley, only mentioning the valley in this short, dismal passage: "Beyond and below lay Owen's Valley, walled in by the barren Inyo chain, and afar, under a pale sad sky, lengthened leagues and leagues of lifeless desert."[3] A month after King's ascent, Muir made the climb. On his first try, he was forced to spend a freezing October night just below the summit. Despite what must have been for most people an unbearable cold, he later noted: "the view of the stars and of the dawn on the desert was abundant compensation for all that." Clearly the view to Owens Valley was more moving for Muir than it had been for King. Muir also makes an effort to diminish the mountain's status as the tallest by adding that "though Mt. Whitney is a few hundred feet higher than Tyndall, the views obtained from its summit are not more interesting."[4]

Anna Mills and three female companions were the first women on record to climb Mount Whitney, making the ascent in the summer of 1878. Clothed in long dresses, with Mills suffering a back injury, these women reached the summit with little difficulty. The last bit of the ascent, which King had considered most harrowing, was remembered by Mills less dramatically: "We passed over the snow-belt, about an eighth of a mile through, with ease, and from there on had no trouble in gaining the summit." She is the first to mention the human marks of the Owens Valley in her description of the eastward view: "Still farther east lay Owens River Valley, with its sparkling lake, winding river, and golden fields of grain. Every road and trail could be plainly seen, and, looking through the glasses, we could see the buildings at Lone Pine and Independence." These early descriptions reveal the significance of the summit view in the experience of climbing the mountain; Mount Whitney becomes the quintessential point of prospect. Merely looking out from the peak rejuvenates the climber, as Mills put it: "For the time being I forgot that I ever was tired; one glance was enough to compensate for all the trials of the trip."[5] As it had done for Muir, the view consoled her.

In 1903 the Sierra Club made an outing into the Kern Canyon with a side trip to the top of Mount Whitney. One hundred thirty-nine people reached the summit. Edward Parsons, a photographer, recorded the trip and wrote of the climb:

> The long line in single file strung along the trail, and repartee and badinage enlivened the tramp. The morning was invigoratingly cool, and as we progressed to higher elevations the coloring and detail of distant ranges, as well as the nearer lake, ridge, cliff, and crag, came out into distinct view until sunlight burst forth on the far Kaweahs and the Sawtooth group to the westward. The ascent, while toilsome, was not dangerous nor difficult, and we made frequent short halts to enjoy the magnificent alpine scenery . . . We looked out over the great valley to the eastward and the Inyo Range beyond, and plainly

distinguished Lone Pine and the meandering green lines of the river and irrigating systems nearly eleven thousand feet below us.[6]

This account offers a rare image of the Owens Valley and its irrigated fields in the same year that the newly established U.S. Bureau of Reclamation (BOR) began taking stream measurements in the valley. Parsons gives a glimpse of the Owens Valley at the peak of its agricultural development, when the town of Bishop was incorporating, the Tonopah mines in Nevada were a strong market for the valley's farm produce, and locals, unaware of Los Angeles's interest in Owens River water, were courting the BOR for a local irrigation project. These early recollections of the view demonstrate the irony of the Mount Whitney climb and the perspective of the Owens Valley from its summit. Reaching the Whitney summit should be more like King's description; it should somehow bring the climber face-to-face with death, but it doesn't. As Muir noted, "almost any one able to cross a cobblestoned street in a crowd may climb Mt. Whitney." Though he added "soft, succulent people should go the mule way."[7] Being part of the most rugged range of mountains in the Sierra Nevada, the peak should offer a Muir-preferred panorama of pristine wilderness, but it doesn't. Mills describes the tiny human scratchings on the land, the beginnings of visible settlement, even from the loftiest of perspectives. By 1878, in this remote sliver of the West, the nation was changing the face of the land. By 1903, it was bucolic. The view, particularly of the Owens Valley, has changed since these early ascents, but the irony of the relatively easy climb and the sublimity of the view still draw people to the mountain.

JULY 10. Evening at Trail Camp, elevation 12,000 feet. Made it the six miles and 4,000 feet up from Whitney Portal, exhausted. Near

11,000 feet the air felt so thin and trees dwindled to knarled and infrequent dwarfs. Above the timberline I stopped countless times, and each pause to catch my breath strengthened my resolve to always live among trees, where the air is heavy enough to fill my lungs with oxygen. My husband, a desert rat on his first backpacking trip, is more at ease in the empty air above the timberline. Views of the Owens Valley from the trail are incredible. The valley floor, seen through the sharp wedge of steep canyon holding Lone Pine Creek, is painted and mottled by cloud drifts. The brown landscape turns almost to liquid shiny mud in its flatness when cast against the narrow granite frame. Flowers bloom all along the trail—lupines, paintbrush, columbine, and shooting star—adding a delicate softness to the granitic immensity of the slopes and cliffs. The air fills with sounds of falling water cascading down the worn granite ribbon of Lone Pine Creek. Thinned and weather-worn pines seem to flow downstream with the water and pool at the Whitney Portal Trailhead, where the terrain becomes less steep, turning it emerald green. At Lone Pine Lake stumpy and stout foxtail pine circle the water's edge like a meeting of old men seated around a Jens Jensen council ring, while the still pool reflects thunder, peace, and granite. So much granite. At 12,000 feet, camp is sited in a cirque, a rounded bowl carved by glacial ice, surrounded by treeless, craggy, scarred peaks. A small lake pools in the bowl's bottom. All around . . . granite, blond as ochre mixed with milk, still and silent. Clouds billowing white and threatening thunder shift across the sky. As the afternoon sun drops low along the peaks, cloud breaks spill heavy rain through thunder claps.

JULY 11. Morning at camp, readying for the hike. Last night, sleeping cradled in granite under frozen air thick with stars, I was awoken by thunderous rockfall, a booming reminder that the seemingly immutable mountains move ever.

Later that morning on the trail to the summit . . . From camp the trickling wet trail up a seeming cliff to the narrow ridge of the needles takes ninety-nine switches, each drifted with snow and blue sky pilot blossoms. Lichens and sturdy minute alpine flowers dapple the granite boulders in fluorescent yellow, orange, olive green, blue, and gold. At the top of the switches, the trail turns and winds along a thin wedge in the sky. To the west lie the gentle forested and meadowed Kern Plateau and emerald glacial Hitchcock Lakes. The ground is scraped and polished as it levels out into meadowlands. Beyond, the Kern Canyon drops deep and thick with green conifers, a wooded feminine cleavage against the surrounding treeless granite peaks. To the east is the frightening drop to Owens Valley, framed by two vertical spires, a line of white billowing clouds, and the tiny wedge on which I stand. I feel as small and precarious as the ladybug that lands on a nearby rock; we linger awhile.

On the summit . . . Reached the summit around noon. Twenty other hikers, including a family of Whitneys from east of the Mississippi who've come to watch their uncle reach the summit for the twenty-third consecutive year, a trio playing whiffle ball, a chihuahua, and a man with a mobile phone trying to call his dad. Everyone takes turns stepping on the highest USGS marker—there are four, since the rocks shift with time. Talk is chatty, mostly of the big, juicy burger everyone seems simultaneously to crave. A man stands apart from the crowd, seeming annoyed that such ordinary people have invaded the solitude of his wilderness experience. The grandeur of the view has humbled everyone down to the most basic human feelings and actions. Men share a bottle of Jaegermister and contemplate urinating over the edge.

The valley opens wide in front of me, now seeming immutable in its remoteness, a patchwork painting in dusty brown. Only in nearness does a landscape come to life with movement, or so it seems as the rocky mound that forms Whitney's helmet-shaped summit shifts and tumbles around me while distance paints the valley motionless. This illusion of flux and fix erodes when high cotton clouds cast shadows onto the faraway flat land, making it ripple with change. They wrap around the brown hummocky Inyo, mottling the mottled hills. Lone Pine Creek wiggles green from the canyon, through the Alabama Hills, which look like melting ice cream, and into the town of Lone Pine. The town reads like a cluster of green between the brown mounded Alabamas and the Owens River, which ribbons thin and green into Owens Lake. The salty dry lake seems huge, plugging the valley's southern end. From so high a perspective the view is no longer reined in by the Inyo and spills over the Panamint Range and beyond Death Valley.

Thunder rumbles off to the west, where a summer afternoon storm has bumped against the high eastern Sierra Range. The highest place in the lower forty-eight is not a good spot to sit during a lightning storm, so we head back to camp after signing the guest book beside the old stone science shed. The mountains take my breath away and don't give it back until I turn downhill toward home. Even the first step in the descent comes easily, each breath bringing the promise of thick air below the timberline, and my whole body sighs in relief to be walking toward the trees.

JULY 12. On the trail, in the trees. Strangers passing on the trail stop and ask how It was, not needing to qualify what "It" is. Single word replies suffice . . . wonderful, beautiful, incredible . . . I savor each breath, not wanting to spill it out on too much idle talk. In this brief passing the mountain is a shared experience, one of anticipation for the stranger, one of completion for me. By these quiet exchanges the mountain creates community in a varied pack of strangers. Looking back toward Whitney, the summit seems so far away, hardly a place where I sat and ate jerky just yesterday afternoon.

Because of the worldwide popularity of climbing Mount Whitney, quotas have now been set to minimize human impacts on the fragile alpine ecosystem. The U.S. Forest Service limits the Whitney Portal Trail to fifty overnight hikers per day from May to October and usually fills the quotas in the first mailbag to reach the district station on April first, when the lottery begins. Those who don't win a nights rest at Trail Camp may try to hike the twenty-two mile, 6,131-foot climb in a day; only about half who try a one-day climb make the summit, because the severe altitude change leaves them breathless. Some take the back way to the top via Cottonwood Pass, the original route of the first climbing parties, but most enter from the Whitney Portal Trail. On holidays, people stream up and down the path—"It's like a freeway" one of the Whitney Station rangers remarked when I inquired about climbing the mountain.

Beyond offering the conquest of alpine peaks, the Sierra has been a cornucopia of natural resources for valley dwellers. During summers, the meadows of the Kern Plateau have served as grazing land for cattle and, at one time, sheep. Formed by the siltation of old glacial lakes, the meadows not only offer nutritious feed for livestock (my great-grandfather swore the grass of the Kern Plateau had extra vitamins that gave his horses pep—"It's like feeding them oats," he'd say), they are also ancient and gentle moments of calm against the austere granite terrain. Cow camps edge the larger meadows, small cabins of rough-hewn lumber, an outhouse and outdoor shaving area for the cowhands, a fence. Old defunct phone lines run from camp to camp, hung on trees through white ceramic transformers. These human marks on the land link the mountains to the valley. They imbue the landscape with memory, with history. With these remnants, the passing of time, no longer measured only by mountain movements, becomes human, comprehensible, fragile, and precious.

Traveling on horseback, the packer leading a string of mules, we reached these meadows from Owens Valley by way of Olancha

Pass along the Jordan Trail, which stretches up one of the many canyons that crease the Sierra Range. This steep climb out of the desert is also used to bring stock up onto the Kern Plateau in the summer.

JULY 28. Long Stringer Meadow. The meadow's sides are lined with granite boulders and glacial erratics—smooth, round stones dropped from ice blocks and named for having wandered from their motherstones—sit in the center of this wash of grass and wildflowers. The air smells thick with life and drones with insects and the occasional hummingbird. Yarrow hovers and twinkles like stars in a vast universe of green. Grass seed heads create a diaphanous haze, the Milky Way, in this celestial meadow. Stillness and quiet make thunderous every nearby sound, the constant chewing of my horse, soft gurgling of the nearby stream, and bird song.

JULY 29. Dawn at Ramshaw Meadow. The sun hits the beaver pond and dewy grass forms fog, quiet and haunting. Shadows stretch long across the meadow. No sound at day-break save the far-off ring of the bell hanging from our lead mule and the nearby gurgling stream. As the sun crests over the eastern mountains, birds join in and the bell tolls louder as the stock come in for their morning grain. Distant cattle groan, perhaps to greet the day. The meadow shifts from a rich mixture of green to golden as the bunch grass, heavy with dew, reflects its straw colors, making it emerge more significantly from the surrounding low sagebrush. Dawn is a fleeting dance of light on water, dew, and mist.

The day in camp begins with putting on the feedbags, listening to the stock's rythmic chewing of oats while I brew coffee. The horses and mules eat oats every morning, an enticement to draw them off the meadows and into camp. One mule's entire burden is to carry the grain needed for the stock. Mules, more than horses, remind me of dogs in their devotion to routine and enthusiasm for simple pleasures. After breakfast we break camp. Food and gear are packed into leather panniers so that all mule loads are balanced to avoid straining their backs. I remember watching my uncle put rocks in the pack bags to even out the weight and thinking it an extra burden for the already laden animals. The sawbuck saddles have two sets of wooden braces, called hanger bars, that cross along the animal's spine. The handles of the panniers are hung on these bars. The saddle is secured by a series of straps and metal rings laced across the mule's belly, chest, and rear end. This lattice keeps the load from shifting about. Once the bags are loaded and other sundry bits of camp gear are stacked on top, large canvas tarps are fastened down with rope to protect the contents from rain and the ever-present trail dust. This routine takes an hour or so each morning and is repeated in reverse each evening when we reach a new camp. Throughout the process the packer, Jim, gently cajoles the mules, promising only to add "just one more thing" to the load. Today we head past Tunnel, Groundhog, Volcano, and Little Whitney Meadows, then descend along the Golden Trout Creek into the Kern Canyon.

Notes from my breast pocket. Each day we spend five to six hours in the saddle, moving along the edge of meadows so they aren't unnecessarily trodden. Jim mounts Clyde, a tall riding mule, and takes the lead, holding the rope that strings together the pack mules—Art, Burt, Shadow, Auggie, and Linda. Shadow, on her first pack trip and just learning the ropes, carries little more than the saddle Jim won at the Bishop—mule capital of the world—annual Mule Days packer competition. She spooks at every bridge crossing, but seems otherwise acquiescent to her new career. I ride behind Linda, who is stricken with gas, her butt becoming a familiar feature of the trip's scents and scenery. My husband, Joe, brings up the rear on Coco, a solid sure-footed horse.

After three days of riding I've learned to take notes, sketch, photograph, and consult my Sierra natural history guide while in the saddle. As an itinerant recorder of landscapes, I carry an array of compact documenting necessities—water colors, brushes, field guides, a camera for shooting slides, spare camera batteries, licorice, and beef jerky—stored in two feedbags turned inside out and hung on the saddle horn. I keep another camera tucked in my shirt away from dust, and I have a palm-sized spiral notebook in my breast pocket on which I scribble reminders, plant and animal observations, smells, images, and ideas brought out in the calm hours along the trail. One advantage riding has over hiking is I don't have to watch my feet, so I can look at and record more of the streams, mountains, and meadows. Each night I read over the scrawl in my notebook, rewriting words that are too shaky to be deciphered later.

At Tunnel Meadow. This meadow is named for the tunnel built in 1883 to drain the Golden Trout Creek into the south fork of the Kern and bring water to the Kern County farmers, who were being hit by drought. The tunnel collapsed not long after completion, amid rumors that farmers who rely on the Golden Trout flow for irrigation had blown it up. It was converted into a ditch in 1891, which also collapsed within a year. Efforts to mingle these waters were then abandoned. A ranger cabin sits on the ridge that makes the natural divide between the two waterways. The outhouse graffiti reminds me to "put the elf back in self."

Later, passing Volcano Meadow. The stock graze on the lush grass and herbs while I photograph the glacial erratics. I think of the plea for elfishness scribbled in the outhouse and how these meadows are apt habitat for mythic imps. I half-expect to see one sunning on a smooth stone, playing a pan pipe. The meadow, despite its name, is a stark contrast to the volcanic Malpais that follow. The great piles of basalt that make up the Malpais look like slag heaps, reminding me of the burnt land of the Owens Valley. The lava flows run in long black fingers through the forest of pines and incense cedar.

A LAND BETWEEN

Further along the trail, we stop at the Little Whitney Meadow cow camp for lunch. An old man carrying buckets of water strung from a pole resting on his shoulders stops to chat, a welcome respite from his task. He prefers the water from the high end of the meadow for its clean, sweet flavor. He talks of fishing for trout and hunting deer. For twenty years he summered at the cow camp and knows the land from living in it. He has now retired to the other side of the meadow.

The southern Sierra has for centuries been lived in, not always kindly but always as more than a place for urban tourists to escape into wilderness. The meadows have been the life source of this human habitation and use. A few cattlemen still run stock in the meadows, though with difficulty. They must contend both with livestock grazing's legacy of abuse, the near denuding of the meadows a century ago, and with today's urban misperceptions of cattle. At a cow camp, I talked over a fence with a hesitant Owens Valley cattleman who still spends the summer season driving stock on the Kern Plateau. He only agreed to chat after he learned that Henry Brown, a man whose presence in the southern Sierra is as deeply rooted as the ponderosa pine that drape the Kern Canyon, was my kin, but he remained wary, worried anything he said would be used against him in the court of public opinion. Whether cattle should graze on public lands has become so common a debate that even those who know nothing of driving stock or have never visited these remote meadows have an opinion on the matter. Grazing practices have changed over time. This cattleman rotates a small number of head from meadow to meadow in keeping with the Forest Service regulations. The task is neither easy nor lucrative, yet he continues; it's what he knows and it brings him into remote areas of the mountains. He and his young daughter were the only people we saw in two days of riding. The quiet, ancient calm of the meadows is reason enough to come when profit no longer justifies the effort. As my uncle Henry put it: "We can only guess at how much of their business [is] economically inspired and how much [is] a cover to go where they wanted to be anyway."[8]

The mountains are, have always been, life-giving to the Owens Valley. Each summer cattle herd upwards in ever-decreasing numbers, while in the spring the canyon streams carry alluvium, the mountain rubble, down to fan out around the foothills. Prior to cattle grazing on the high mountain meadows, the canyon trails were used as trade routes by the Numu and neighboring native communities. This exchange between mountain and valley has gone on for centuries, transferring nutrients, water, salt, and earth by way of water, cow, human, and other natural forces. The imbalance in the exchange is revealed in the heaped debris that has formed a belt along the length of the eastern range of the Sierra Nevada.

At the base of this alluvial belt below Mount Whitney, a cluster of stones form the Alabama Hills. The hills borrow their name from nearby mining claims staked in 1863 by rebel sympathizers. The owners were honoring the ship that sunk the Union ship *Hatteras*. A year later, when Union supporters discovered gold to the north of the Alabama claim, they retaliated by naming their claim *Kearsarge*, after the ship that had just sunk the *Alabama*. Both names lasted in the valley long after the mines had been abandoned, and they preserve hints of local tension brought on by the Civil War. The Alabama Hills' fractured and softly weathered rocks conjure images of pagan landscapes such as Carnac or Stonehenge. The rocks are stunning mounded globs of granite cracked and fissured every which way and tinted with orange and green lichen. The granite wears a weathered cream skin that blotches black on older patches not yet sloughed off by frost heave. Hanging bits of vegetation nestle in cracks between the stones. Once thought to be ancient in geologic origin, the Alabamas are actually part of the same granitic block that formed the Sierra Nevada —so they are the same age as the craggy, younger-looking Sierra. Their different character is the result of differing weathering conditions; the high mountains are sculpted by glaciers and an arctic climate, while the Alabamas in the valley below have been more gently shaped by dry desert winds, rain, and frost heave.

Movie Lane, so called for the many movies filmed in these strange hills, meanders among the rocks. One can almost hear the clang of grub pots in some John Wayne Western. Or Walter Brennan, playing that old geezer cook going on about gold or cattle and cackling to himself as he stirs a big batch of beans. Or the rifle ricochet from the endless gunfights in Clint Eastwood's *Joe Kidd* (1972). Films made in and around the Alabamas celebrate genuine cowboys—those smelly, whiskey-drinking men who have few lines so they read 'em real slow. The Duke, who probably did more to romanticize the cowboy and his Western landscape than real cowboys did, still haunts these mystic stones. But this is also the land of the Lone Ranger, the place of legendary men who rode horses, carried six-shooters, and never got dirty; the television cowboys decked out in hankies, stretch pants, and spurs.

Farther east across the valley lies the White-Inyo Range. All that the Sierra possess in grandeur and spectacle, elegantly clothed year-round in snow, these mountains hold in humble nudity. Forming the eastern edge of the valley, they lay brown and almost treeless when held against the Sierra Range. The White-Inyo Range runs north to south like a petrified caterpillar and is a folded mixture of the three major rock types—igneous, sedimentary, and metamorphic. The oldest rock is the siltstone and limestone of the Wyman Formation, seafloor sediment deposited between 2.7 billion and 700 million years ago. From the air, the orientation typical of basin and range country associates the range with the multitude of other caterpillar mountains that, in the rippled heat of summer, seem to wriggle across the desert towards Mexico. The range is artificially divided in name along the midsection at Westgard Pass—the White Mountains lie to the north and the Inyo lie to the south of the pass. Appearing dwarfed by the Sierra, an illusion caused by the desert basin's trick of lying flat and making all distances incalculable, the Inyo bridge the gap between the monumental Sierra and the small human settlements scattered along the narrow ribbon between these mountain ranges. Most settlements and roads are west of the river, closer to the streams that come down from the Sierra canyons and closer to the Sierra Range. So, from the towns and villages that hug the valley's western edge, the Inyo Mountains seem to shrink away across the valley. Around Bishop, fifty miles north of the Alabamas, the valley widens, the Sierra peaks recede, and the White Mountains rise high enough to be glaciated. There, at the widest part of the valley, the contrast between the ranges wanes. And northward of Bishop is the highest peak in the range, White Mountain (elevation 14,246 feet), only 250 feet shorter than Mount Whitney.

At the drier southern end of the valley, volcanic cones edge the foothills of this smooth and blackened range, which is, as Mary Austin described, "burned, squeezed up out of chaos, chrome and vermilion painted, aspiring to the snow-line."[9] Here the naked desert is at its best, tipped upwards for all to see. The White-Inyo Range is in the rain shadow of the Sierra, so it gets less moisture, an effect that can be gauged in the alluvial fans that open out of each canyon. Less debris flows off these dry mountains, so the fans remain distinct rather than coalescing to form an alluvial belt like the one that skirts the Sierra Range. Newcomers to the valley may drop their jaws at first sight of the Sierra, but it is the union of both ranges that turns initial awe to sustained admiration. The juxtaposition of these geologically varied ranges creates a resonance in the land between, like the throbbing tines of a tuning fork. Both ranges are massive enough that their gravitational pull is almost tangible. The landscape vibrates with tension as the mountains yearn for one another.

At dusk the Inyo rest in soft light shadowed only by remnant storm clouds, their bumpy, treeless sides reminiscent of an aging woman slumping toward her grave, her once-jubilant youth sagging with living, childbearing, and time. Inyo beauty is an old woman's beauty, revealing itself in life's tragedies, joys, loses, and loves, recorded in every fold and wrinkle of her aging flesh. The mountains are dusty and naked, well-patinated skin whose years of touching compel you to touch, to feel the warmth left from past caresses. It is perhaps this warm human feel of the Inyo Mountains that make them and not the Sierra "the dwelling place of a Great Spirit" for the Numu.[10]

The White-Inyo, like the Sierra, has been a place of harvest for valley dwellers. The pinyon pine, whose nuts were gathered each fall by the Numu, grow most abundantly in the White Mountains between 6,500 and 8,500 feet. Near mining operations, the trees were harvested as fuel wood for smelters during the fleeting silver-mining boom of the late nineteenth century. The lower naked mountain slopes, easily read by mineral prospectors, promised rich stores of ore. But the dry hills invite little settlement because the sparse streams and springs offer no reliable water supply, especially when compared to the abundant Sierra creeks. Consequently the White-Inyo's western flank only hosted visitors who had come to extract some resource such as gold, silver, lead, zinc, or talc from its aging sides. Even the mining towns such as Cerro Gordo, site of the largest silver strike in the valley, were more like long-term guests camps than settlements.

These mining ventures have left the mountainsides scarred

with oozing cavities and cluttered with abandoned mining equipment scattered at their feet. An occasional military plane, probably heading home to the neighboring China Lake Naval Weapons Center, hugs the lower elevation of the Inyo near pockets of the dolomite mines. Along the slope, dolomite tailings drip like white chocolate, leftovers from what was hauled off to decorate the Mills Building in San Francisco, provide flooring for the Los Angeles International Airport, or serve a less illustrious fate as the chalky crushed rock sprinkled on roofs in suburbia.

Tucked between these ranges lies the Owens Valley, what Austin called a "land of lost rivers, with little in it to love."[11] Taken together, it is a scene of great beauty. Despite tremendous change along the valley floor since 1800, the constant presence of the mountains creates a landscape of enduring power and splendor. W. Newton Price, an old-timer who lived his life between the mountains, recollects his father's first view of the valley, of how the land in its configuration had the ability to make one feel at once humble and small, yet connected to something powerful. "As many times as I have been privileged to drink in this scene, each time I am gripped by the same emotion which makes me want to shout, 'Of this greatness and vastness, though as a pebble on the face of a mountain, I am part of it!'"[12] This feeling echoes other descriptions of spiritual awakening. It is to sense with humility one's humanness, the mortality and imperfection of it all, yet feel a part of something divine.

MARCH 11. On the road to Onion Valley. The sun dips beyond the Sierra . . . Storm broken, revealing stunning snow-capped mountains. Driving up the road again toward Onion Valley—stop and start, unable to move the rental car through the landscape. God floods my thoughts as I look out to these mountains. Geologic upthrusting, faulting and climatic conditions for alpine misting may explain the physical phenomenon that lies before me, but divinity is at work that lets me perceive beauty, heart-wrenching, car-stopping beauty. My ears ring in silence, only the swooshy sound, like wind on crows wings, as the late-day air squeezes between Kearsarge and Independence peaks. Northward a ghostly gray fog masks the mountains. Sunshine glows against the range's far southern end as it tips a bit towards the valley floor. From both north and south the White-Inyo and Sierra seem to bend toward each other, like lovers from across the room.

When a storm drifts into the valley from the west, it first climbs the saw-toothed ridge, where many squalls are stopped before they can spill over into the valley. The mountains steal their thunder, leaving them still fierce but immobile cloudbursts high amongst the granite peaks. Stronger fronts overpower the mountains, shrouding them in a cloak of whiteness. From the valley this is a spectacular event to watch, a battle between impenetrable stone and diaphanous mists. When the mists win, these massive mountains fade to white, only their toes peaking from behind the cloud curtain. From a distance it looks like a gentle siren's seduction; the vapors seem soft and delicate. Storms often rage into the night, rattling windows, settling dust in forgotten fields. Rain may fall in the bottomlands, wind-whipped water shaken from the battle-worn elements, but the struggle remains hidden. Only daybreak reveals the remnants of this brawl. Unlike Midwestern tornadoes that leave trails of broken landscape heaped with debris, when the storm mists dissipate here only whitened peaks and the blinding brilliance of freshly fallen snow remain. And so the valley's water cycle begins.

The morning after the Whitney hike I awoke to such a scene, a rare event in summer. Beyond my window at the Dow Hotel, a double rainbow spanned a dewy green field in Lone Pine. Moisture filled the air with electricity and refracted light, making each chilled breath taste of light and mountains. So tangible was the air, this delicious air, I almost grasped for it, as though it could be held and taken home. I thought back to my childhood, of the sand crabs I used to sneak into the car after a day at the beach, of my impulse to take a piece of the place with me, to make the pleasure last. But like the stinking dead crabs my father would find days later, tucked under the car seats, this air would lose its appeal. It is best left in lungs, in memory, and lightly lingering on the tongue.

The valley's hydrologic patterns rely little on local rainfall. Casting a formidable rain shadow across the valley, the Sierra blocks all but a scant few inches of rain. Once it crests the high eastern range, the adiabatically dried air, whose moisture is stripped during the eastward march across the Sierra, sucks any dampness from the valley. The bottomlands might wrest out only five to six inches of rain in a year from passing storms. Mountain snowfall from autumn into late spring holds the valley's water, which is released when the warm winds that beckon summer blow in from the Pacific and melt the snow off the mountains. Occasional moist tropical air moves in during the winter months, sometimes causing an early melt and flooding, but the bulk of the snow is held fast in the Sierra well into late spring. Among the alpine peaks, small patches of snow cling on through summer, turning pink with algal blooms.

Cascading down steep ragged ravines, snowmelt-fed creeks roar, heft boulders, break trees, and fill the air with spray and thunder. Each heavy flood resculpts the sharp forested terrain, deepening the clefts between granite peaks. The creeks lose their ferocity once they are no longer pinned in by granite, and they then begin to let loose their alluvium, dropping heaviest boulders first and carrying pebbles and sand far out in a fan pattern. Each season the slope steepens and sediment pushes out across the valley floor. Long ago, these alluvial fans joined to form a uniform wedge of debris along the feet of the eastern Sierra Range. Now each creek snakes across this bajada and, ribboned with willows, meanders toward the bottomlands. They once trickled through wide meadows, flooding them in spring and turning them to marshy bogs. What didn't linger in a lowland meadow fed the Owens River. Its banks thick with cottonwoods and willows, this perennial river flowed southward into Owens Lake. Moving through the flat valley, the river bent and wiggled like a worm, and, having neither an outlet to the sea nor enough volume to breach the volcanic flow blocking the way to lower elevations in the Mojave Desert, the water came to a final resting place at the saline Owens Lake.

For centuries the valley landscape went through these seasonal hydrologic fluctuations in balance with vegetation, wildlife, and the few human inhabitants. As the river swelled, sediments held in the flow scarified and distributed seeds and scoured clogging reeds and rushes from the stream channel. As summer progressed, the flow subsided, reeds returned, and seedlings put down roots to ride out the next spring flood. Throughout the seasons ample groundwater sustained a dense band of riparian vegetation meandering through the desert valley floor. This was in no way a passive ecosystem, with its yearly tantrum and violent churnings, but it was a natural one.

Native peoples who populated the valley began to change the natural annual water cycle. No one knows when the hand-dug canals that diverted water from streams to irrigate meadows for the Numu were built, but they were present at least as early as 190 years ago. The canal system, developed mainly in the valley's northern extent, included miles of ditches and many small earthen dams. The system watered meadows of wild plants such as *tupusi*, yellow nut-grass, which was one of the most important food sources for the Numu. The ditches were later used to irrigate the tilled fields of the pioneers. In the West, mixing water with people stirs up a thick political brew; this is especially true in the Owens Valley. Of all the features in the valley, water has rallied the greatest ongoing debate and controversy, and, even more than mineral prospects, water extraction has transformed the Owens Valley. When European, American, and Mexican settlers first entered this narrow valley, their strongest impressions were of the great snowy water stores in the high Sierra Range and the lush marshes in the bottomlands. Almost without exception, every personal account of the valley includes discussion of water, which is not so surprising since available fresh water made this arid valley habitable. Owens Valley was the jewel of the region, as illustrated in John Glanville Dixon's description of his parents' first impression of the valley:

> To the right, the snow-crowned glory of the rugged Sierra Nevadas was spreading a silent benediction of shadows across the valley floor and up the sides of the Inyo Range. They stood there transfixed, speechless, for this, they knew, must be the beautiful Owens Valley that Brother Will had written about, and it was love at first sight! When at last Galen could speak, he said, "You know, Sadie, the gold, silver, and other mines of the West are gradually being worked out, leaving of coal mines and oil wells will also gradually disappear, but the 'white gold' of the snowy Sierra, through the foresight of a kind Creator, is annually replaced." Pointing to the snow-capped range, he said, "There is California's *real* Mother Lode, a wealth of water, which cannot be taken away."[13]

Arriving with horses, mules, oxen, and cattle, nineteenth-century emigrants would undoubtedly head for the meadows so their stock could graze. From this vantage point they might be able to forget they were in a desert. It would be easy, especially with the pioneer tendency to see what could be rather than what is, to imagine a fertile landscape waiting for its green pastures and rich soil to be cultivated by industrious young emigrants. Water was their ticket to freedom, "the secret of their happiness," the magic element that would enable them to play alchemist and turn what was in fact a sagebrush expanse into thriving fields and flowering orchards.[14] The *Sierra Magazine* offers the most optimistic perspective of the Owens Valley for settlers and miners in its commentary on water: "This is a peculiar land, and the soil has peculiar qualities. It requires water. Give it water and you work a miracle. Put an abundance of moisture on this sand and the sand disappears. In its place comes a rich loam. The ground is full of nitrates. It is full of potash, It has all the plant-life-giving characteristics a farmer most desires and it has them in inexhaustible quantities."[15] Although the farmers, ranchers, and miners diverted more water than the Numu had done with their canals, no one in the valley had the chance to test the inexhaustible nature of the desert soil. Time has shown that this annually replenished white gold doesn't come with a guarantee. Between climate fluctuations, which have decreased snowfall in the Sierra, and the city of Los Angeles, which now appropriates most of the snowmelt that runs down the mountains, the valley's elixir no longer works magic on the land.

Natural events that have altered the hydrology in the valley are both climatic and geomorphic. On a geologic time scale, the valley has been growing more arid since the Pleistocene Epoch. Owens Lake once connected with other basins. After breaching the lava flow at the south end of Owens Lake, the water flowed into the Mojave Desert and the Indian Wells Valley, pooling in China and Searles Lakes. When the climate was at its wettest, the water may have continued east to Death Valley. What remains of this Pleistocene lake system are salt-encrusted playas that only shimmer after a hard rain. The 1872 earthquake centered near Lone Pine, thought to be one of the largest in California history and registering around 8.3 on the Richter scale, toppled many adobe homes and killed twenty-seven people. It also fractured water flow from the Sierra to Owens River along the shifted fault zone and formed a small reservoir called Diaz Lake. After the quake, without the life-giving Sierra waters, the river's riparian vegetation declined, and it has further suffered since the Los Angeles diversion, in 1913. Through natural and manipulated events, the valley is giving way to the desert, making water precious by its scarcity.

Wallace Stegner, in *The American West as Living Space* (1987), argues that aridity, the absence of water, is what defines the West; it is the creator of Western physical and social form.

> Aridity, and aridity alone, makes the various Wests one. The distinctive western plants and animals, the hard clarity (before power plants and metropolitan traffic altered it) of the western air, the look and location of western towns, the empty spaces that separate them, the way farms and ranches are either densely concentrated where water is plentiful or widely scattered where it is scarce, the pervasive presence of the federal government as landowner and land manager, the even more noticeable federal presence as dam builder and water broker, the snarling states'-rights and antifederal feelings whose burden Bernard DeVoto once characterized in a sentence—"Get out and give us more money"—those are all consequences, and by no means all the consequences, of aridity.[16]

Human intervention affects the landscape profoundly, with dams, aqueducts, and irrigation—all responses to aridity—in the West's desert reaches. Aridity has been considered a problem to be solved, the kind of surmountable challenge that has made the young country swell with pride at its own accomplishments and made the fruit of the irrigated harvest taste sweeter. Some might argue that the highway system is the larger landscape transformer in the West, but a drive through California's Central Valley on Interstate 5—a narrow asphalt strip floating in a sea of earth exhausted from irrigation—shows that aridity, not mobility, has been the greater catalyst for landscape change in the arid and semiarid West.

This pattern is echoed in the history of the Owens Valley, but with a twist; rather than, as in Stegner's description, being conquered by dam-building engineers, aridity has become more pronounced. Where the arid desert landscape once lived in balance with soggy meadows along the bottomlands, it now reigns triumphant, sending in troops of sagebrush, saltbush, and rabbitbrush to dominate the drying marshes. Since 1913, when river water began to drain into the Los Angeles aqueduct, the desert has reclaimed the Owens Valley. Before diversion, the Owens River's annual discharge measured 300,000 acre-feet (98 billion gallons) per year, enough to fill thirteen million backyard pools or to support two million Southern California residents. The Los Angeles Department of Water and Power (DWP) counts this flow as a savings, as it is water that would otherwise, from the DWP's perspective, be wasted in the saline Owens Lake.

The lake once supported a huge population of picawada, fly pupae that supplied protein for the native people of the southern end of the valley, who didn't have extensive pinyon woodlands nearby. Eaten as a mush, the pupae were an important food source, though said to be less tasty than the pupae in Mono Lake to the north. In the pioneer days of the late nineteenth century, the lake water was shipped to San Francisco and "sold as a specific for about every ailment known to quackery or science."[17] By 1928 the lake had evaporated, and it is now an alkali sink, a giant dull pink eye at the end of the river's reduced trickle, where trucks carrying loads of mined salts across its crusty surface look miniscule. The salts left behind during evaporation have accumulated and are now nine feet thick on the exposed lake bottom. Up to 500,000 tons of trona, a natural source of sodium carbonate, is mined each year from the lakebed for the production of glass and detergent.

Currently, only the occasional rainstorms that wet the surface make mountain reflections dance on the thin film of moisture, as they had when the lake held cool, blue water. The moisture from such rainstorms fixes the salt to the playa, but more often, when wind blows across the dry lake, the air is filled with a white cloud of dust containing sodium, silicon, sulfate, and arsenic. The PM-10 emissions, particulate matter of less than ten microns, are monitored because of their threat to human health. Owens Lake produces 6 percent of America's dust, half a million metric tons each year, and its clouds of PM-10 dust travel as far as the San Gabriel Mountains, near Los Angeles. A court decision in 1995 gave Los Angeles the task of keeping the dust down. The city must follow the 1998 plan developed by the Great Basin Unified Air Pollution Control District to reduce PM-10 emissions. The plan mandates the use of shallow flooding, managed vegetation, and gravel cover as measures to reduce dust on nearly twenty square miles of the lake's worst dust-producing areas.

The river, dammed in three places and drained into the aqueduct, has also lost its natural vigor. In 1862, Lieutenant Colonel George S. Evans, who had been sent to the valley to "chastise" the Numu, swam the swollen Owens River. He crossed the river to meet with Captain Rowe, who was endeavoring to make peace with the native peoples of the region. Though Evans outranked Rowe, the Captain's health was poor, so he couldn't cross the river to report to Evans. They had tried to talk across the water but were unable; Evans' report in a letter dated July 9, 1862, describes the event: "I immediately saddled my horse and rode down to the river, and finding it almost impossible to talk from bank to bank in consequence of the sloughs on either side of the river being swimming, I resolved to cross myself. After swimming two sloughs and the river and wading half a mile through willows and tulles, I reached the eastern bank of the Owen's River, where Captain Rowe was camped, and spent the night with him."[18] By July the snow-melt waters had breached the river's channel, spilling into the wide marshland that once skirted the riverbanks. Today one could talk in whispers across the waters of the Owens. Since the aqueduct diversion, the tules have receded to meager patches beside still pools in the narrow river.

MARCH 13. On State Route 136 at bridge #48-2. Tules lay matted on the river like thick and loose piles of hay, the water lost under these thirsty plants, buried as though dead. Muted sounds break the cold morning silence, a trickle under the dried reeds, and the gurgle of a startled frog. Aging broken cottonwoods rest along the banks leafless, lifeless, and still. Slender green shoots of salt cedar, one leafing branch on a fallen willow, the only green along the river's edge. The water's ink black surface reflects the chilled sky and wisps of blue breaking through the wintry clouds. A killdeer says its name over and over, a meadowlark sings, a crow calls. No young cotton-woods have rooted beside the river in years. When the last of these rotting trees collapse along the banks, nowhere will the birds sit to greet the day; even a dead tree offers a perch.

After diversion the valley's plant communities began to adjust to the increasingly dry conditions—riparian vegetation died back, meadows disappeared, and desert scrub advanced onto once-watered land. Plants are great indicators of environmental change, bound as they are to the land. Without the ability to move about, except for rogues like the tumbleweed, they grow in places suited for them. Changes in a plant's range can reveal underlying shifts in climate and soil conditions. Take for example the conifers that now lay as petrified stumps strewn about the Painted Desert in Arizona or the bee pasture of California's Central Valley, described by John Muir as "one smooth, continuous bed of honey-bloom, so marvelously rich that, in walking from one end of it to the other, a distance of more than 400 miles, your foot would press about a hundred flowers at every step."[19] The forests died out as the North American continent drifted and global temperatures shifted, and in much of the Central Valley, where soil is exhausted and salinated from irrigation farming, much of the land can now only sustain salt-tolerant crops like cotton. The Owens Valley vegetation has similarly changed in response to natural influences and artifice.

Thankfully, the valley's diverse landforms create countless microclimates, varied soil conditions, and habitats for many different plant species, so, despite changes in the past century, the valley and mountains still offer a feast for plant enthusiasts. Moisture-loving species collect along stream corridors and in marshland, while water-conserving plants grow in the drier bottomlands and bajadas. Vegetation less tolerant of heat climbs up canyons, seeking shade and cool temperatures. Most lowland species are alkali tolerant, thriving in the salty soil beside the saline playa. And among these myriad natural vegetation communities lie planted alfalfa fields, forgotten apple orchards, and cottonwood-lined streets of the agricultural landscape. This diversity of plant communities tucked within the narrow valley gives visitors and dwellers varied impressions of the landscape. The valley has been called a land of milk and honey, a green oasis, a dry barren desert; all fitting names in different places or at different times.

Draping the landscape like a patchwork quilt stitched together along seams of creek, river, and bouldery folds, plants soften the rocky arid land. Their smells lock the landscape into memory for the traveler, farmer, miner, all. Vanilla odors wafting from pinebark cracks, smoky remnants lingering on a twice-worn shirt smelling of last night's sagewood campfire, the damp creek-side wildflowers teeming so thick with life the scents rest on the tongue like food. A friend once remarked while we camped on the southwest desert: "Who needs food out here—you can just eat the air." Like the wind that blew that night across Hovenweep, the Owens Valley air, filled with plant perfumes, could make a meal.

Moving eastward across the valley, I sense the land's complexity best by watching how the vegetation communities change. From near the Mount Whitney summit to the low Sierra foothills, plants act like ambassadors to each small and fragile visitor panting along a trail. Even the tiny alpine flowers bring the landscape down to size and make it comprehensible. The journey begins atop naked granite, sharp craggy rock barren of even tiny alpine blooms. On Mount Whitney all is stone, cracked, oft-frozen and lightning-struck granite. Perhaps below the rubbly surface dusts of soil linger, but it is too far from the frozen sunlight to harbor even the dwarfed plants of the alpine landscape. Marmots and ratty-looking finches scurry among the rocks, mostly scavenging for bits of cheese and crackers dropped by hikers lunching on the peak. The birds beg like naughty dogs. Yet just below the summit two ardent little blooms hold fast in cracks and crevices.

Sky pilot, with its single lollipop ball of blue blossoms, spray the granite with color, as if dots of sky itself had rained down on this monochrome landscape. And like small bursts of sunlight beside these dotted blues, alpine gold turn their yellow faces towards the sun. Among the smallest of the sunflower family, they grow in miniature like their tall relatives harvested for seeds and named in

Italian *giro del sole*, in French *tournesol*, and in Catalan *gira-sol*, for their propensity to perpetually turn their faces to the sun throughout the day like natural timekeepers. With a short four- to seven-week growing period these tiny plants push out only a leaf or two a year. Laying on my belly to observe the blooms up close, I marvel at the effort to create such delicate beauty, far beyond the elevation of even the heartiest tree. All other alpine plants that spot the granite landscape on the descent share the sky pilot's and alpine gold's stature. In a land of frequent bitter cold gales, lightning strikes, and snow, staying low to the ground is a natural survival tactic. Like the desert plants growing below on the valley floor, these tiny alpines often have more going on underground than at the surface, sending out an extensive root system to search for and store nutrients.

Lichens paint the stones along the switchback trail in fluorescent yellow, rusty orange, olive green, and black. These symbiotic combinations of fungus and algae form strong acids that slowly eat away at the granite boulders on which they cling and, over centuries, turn them to sand. Alpine meadows edge the meandering feedwaters of Lone Pine Creek. Matted grasses are splashed with magenta shooting star that encircle stones resting deeply in this ancient garden, tended by climate and time. Shrubs, grasses, and alpine flowers become more abundant below 12,000 feet and trees crop in at about 11,000 feet. Foxtail pine is one of the tree species that can survive the thin air, rock falls, and lightning strikes of high elevations. These timberline trees cling to the canyon walls, looking like old miners beaten down by years spent wrestling a living out of stone. Like the ancient bristlecone pine of the White Mountain peaks north across the valley, over time these trees grow gnarled and barkless.

Almost more stunning by their delicacy, played against the stone cliffs, are the multitude of wildflowers that grace the trail and streamsides. Blooming lupines, paintbrush, columbine, penstemon, wild rose, yarrow, mimulus, shooting stars, delphiniums, and dandelions follow the seasons up the slopes. One can keep a calendar from early spring to midsummer by following flowers up the canyon, finding the last bloomers opening in the alpine fell-fields as late as August. They color the landscape in a mix of reds, pinks, orange, yellows, blues, violet, and white—a rainbow draped across the stony canyon. They brush against my chest as I wind along the narrow path, intoxicated with the fragrance of blossoms and damp acidic soil. Farther down the canyon the air becomes warm and dry, a reminder that it's still the height of summer in the valley. Cicadas scream their usual summer song like background music to a Clint Eastwood Western. The air smells of dust, sun-baked sagebrush, and the vanilla in the Jeffrey pines. Butterflies dance around phylox blooms among low dry bush chinquapin and ceanothus that cover the hot dry slopes.

At the base of Lone Pine Peak, the stream that cascades down the steeper mountain slope carves a canyon in the alluvium. Pinyon pines follow the water out of their usual mountain habitat. Within the wedge of the canyon, they are sheltered from the hot summer sun, though some spill over the canyon wall onto the flat bajada. These hearty plants crouch low, staying in tight, dark green balls to protect themselves from the heat. Mormon tea grows in bright yellow-green clumps on the high slope among granite boulders. Sagebrush dominates on the high bajada, where an occasional cholla cactus pops its head up from the ocean of gray-green foliage. In the spring, dry stalks of last year's bloom hang above the sage like a pink sienna mist over a still bay at dawn. Sagebrush is the plant that ties this narrow ribbon valley to the larger Great Basin Desert. Though not a true sage, its pungent odor after a rain is reminiscent of the herb. The smoky scent of a sagebrush campfire smells like the pages of a Zane Grey novel. It is the quintessential desert cowboy plant, as necessary to conjure up the nostalgic figure as his horse, rifle, or jangling spurs.

MARCH 14. Below Lone Pine Peak. Water, spilling over granite boulders beneath mats of leafless willow, drowns out all sounds save the trill of birds and occasional roar of military jets. Water, the landscape's melodic music even on this drying slope, gurgles and splashes in quiet contrast to the roar made while it crashes down the canyon from the high snowy peaks. Small oaks grow at the water's edge amongst the willow, which are full with spring catkin buds, together creating a scrubby, matty wet thicket. Above the stream's steep bouldered bank, the ground is covered with loose and sandy decomposed granite. Little tufts of bunchgrass grow sparsely; the larger dry clumps shorn short by cattle. Orange, bright yellow-green, and black rings of lichen cling to the boulders. In the shade of a pinyon whose trunk has fused to a granite boulder, a single currant plant grows delicate and sprawling with tiny yellow flowers in early bloom.

In an afternoon cloudburst, the pebbled bajada looks stroked with pigment where different species dominate. Everything is soaked, washed clean, and brilliant. The foothills stretch long and fingery along the slopes like sea slugs groping toward the valley floor. A whippoorwill calls its name. Red willow whips poke above the sage, marking Lone Pine Creek. As I walk back down the slope and the stream gurgle fades, the evergreens thin, giving the landscape over to sage, bitterbrush, and obscenely green Mormon tea. Tufts of buckwheat blooms dot the openings in the sage, giving a gardenesque quality to the arid planting.

Patches of blackbrush move in further down the bajada, becoming more dominant with decreasing elevation. Shrubs of the lower bajada grow shorter, more rounded and compact than the sage. Dried and dead looking, they grow in blackish clumps and prefer the ridges of the fingery mounds that stretch down the alluvium between washes. Among the Alabama Hills, plants become more diminutive and sparse around the granite heaps, and at the base of each sculpted stone soil takes on the rock's ruddy ochre hue. Blond decomposed granite drips down off the Sierra to mingle with the Alabama tints like spilt ice cream on a hot sidewalk. So close to abundant snowy stores of water, this desert landscape lightly clothed in vegetation seems soberly reserved.

Moving through the Alabamas, the planted landscape comes into view. People and streams are easy to spot in the desert, since they're always associated with the tallest vegetation in sight. Big trees—cottonwoods, willows, and pines—at Cuffe's Guest Ranch appear against a rocky hill even before the buildings become visible. Lower still, the rabbitbrush takes over and saltbush appears in the descent through the creek canyon, hinting toward the saline soil of the bottomlands. On the gravelly slopes, thinning vegetation freckles the tawny earth with splashes of yellow carpet, a tiny yellow wildflower beginning an early spring bloom.

The road straightens into the town of Lone Pine, and the stream becomes a tamed trickle running through a clipped park lawn. Today buildings give the rural town a sense of human order, but during the course of settlement planted vegetation helped establish this order. The first farm fences were made of low sod walls planted atop with willow whips, so plants delineated property divisions. Yet now the town's vegetation seems peripheral, almost forgotten amongst buildings and the mountainous Sierra backdrop. Italian cypress, grazing fields, and flower boxes replace the vanished desert, creating a strange mix of the agricultural and ski lodge vernacular. Boxes and bordering beds of annual flowers edge shops and hotels, while evergreens dominate the clustered shrub plantings that ring buildings. Architecture has supplanted vegetation in giving the landscape a human scale. This has not always been so.

Descriptions of the valley in the late nineteenth century are laden with images of a remote and hidden oasis, tucked away behind treacherous mountain ranges. The Numu canals that irrigated the meadows must have increased the contrast between these green grass fields and the surrounding desert scrub. Numu villages were sited beside the meadows. Nineteenth-century emigrants settled on top of these villages, now the town sites of Bishop, Big Pine, Independence, and Lone Pine. As they expanded the existing irrigation system, the desert shrank further from the valley floor and all but disappeared from pioneer memory. Yet, even then, nearly all valley vegetation was desert scrub. In 1863, William H. Brewer observed while exploring for the California State Geological Survey Botanical Department: "The Sierra Nevada catches all the rains and clouds from the west—to the east are deserts—so, of course, this valley sees but little rain, but where streams come down from the Sierra they spread out and great meadows of green grass occur. Tens of thousands of the starving cattle of the state have been driven in here this year, and there is feed for twice as many more. Yet these meadows comprise not over one-tenth of the valley—the rest is desert."[20] Without irrigation, most of the land could not support crops, orchards, or cattle fodder.

The Numu, who dwelt in the valley long before nineteenth-century emigrants came to settle, were a mobile people who traveled extensively within the valley and along established trade routes. They lived by the streams but spent much of their time gathering food among the desert vegetation. Their myths and stories, like Brewer's observations, describe the valley as dry and hot. In contrast, nineteenth-century emigrants were settlers. They established farms, ranches, and towns beside the green boggy places in the valley, where they spent their days working the land. They recall the green and lush valley, as long-time valley resident Elizabeth Carrasco adamantly exclaims: "Before [the Los Angeles DWP] came, the whole valley was full of orchards, lots of wild feed, and lots of all kinds of grain. I have heard stories, in fact, I heard one the other day, that this was a very arid valley when they came here. That is a joke. My dad came here in 1864. He was just a small boy, and they came in over the mountains from Utah. He said when they came over the mountain, it looked like they were coming into Paradise. Everything was green, and the horses and cattle were up to their bellies in feed."[21]

Upon seeing the lush ribbon of riparian woodland that meandered along the valley floor and the green meadows pooled at the base of the bajada, the settlers set out to transform the surrounding scrubland into their version of Eden. Pushing the desert farther out from settlement and replacing it with a planted landscape made the Owens Valley into a verdant paradise. This is how most settlers remember the valley. Since the Los Angeles aqueduct diversion, most rural land has dried and decayed, the orchard trees long ago felled and sold for firewood. The Arcadian buffer no longer holds the desert at bay and, particularly in the southern Owens Valley, only ghostly, forgotten vestiges of the once-rich farmland remain. The bottomland now serves mostly as cattle graze.

MARCH 14. In the bottomlands on Mazourka Canyon Road. I've parked along the road and am looking out across the grazing land, beyond barbed wire and a bull, to a line of green cottonwoods—tall mature holdouts of the rural oasis. Elsewhere golden dry grasses and gray-green rabbitbrush blanket the flat landscape. Towards the river grow clumps of saltbush and sage. Owens River runs under the road in two galvanized pipes, like black tar into a pool, disappearing downstream under a mat of dry tules. All the cottonwoods along its banks, still bound by late winter's chill, look dead.

Cattle bellow to one another across this desert rangeland. Walking narrow paths, like fat ladies in tiny high heels, their enormous bellies surge from side to side like a thick liquid above dainty and nimble feet. They are the landscape managers of the American West, traipsing about selectively weeding vegetation according to their tastes, crapping out seeds here and there to distribute plants. Planting and pruning the desert vegetation into gumdrops with their grazing, cattle have changed the lowland vegetation ever since their introduction in 1859.

Of all the valley and mountain vegetation communities, that of the bottomland seems most abused by human hands. The lowlands have sustained continual use—first by the Numu, then by farmers and ranchers, then by diversion, and now mostly by grazing cattle—and harbor the vegetation that is slowest to recover from disturbances, second only to the alpine miniatures. Growth of lowland desert vegetation is incremental, since desert soil, even in its natural state, supports only slow and sparse growth. Once disturbed, the land can take many years to re-establish vegetation. Plowed fields lain fallow invite species such as rabbitbrush, which thrive in disturbed soil, exacerbating the already difficult struggle for a diverse community of desert plants to re-establish. Yet the desert vegetation patiently returns, weathering many seasons as it slowly overtakes land that is no longer watered. Consequently the southern valley floor now has even more desert vegetation than Brewer described in 1863.

JULY 12. Overlooking the valley from the Inyo at 6,300 feet. I move eastward and the landscape grows more arid in a smooth transition from the grazed bottomlands to the desert scrub in the foothills. Looking out to the valley from the Inyo, quiet hangs in the air so I can hear only the ringing in my ears, cricket calls, and the occasional fly buzzing about my face. No trees grow here—just desert shrubs, such as sagebrush, saltbush, Mormon tea, and buckwheat, sparsely set amongst the granite boulders that are scattered and piled about the rolling landscape like dinosaur eggs. A dry wash shifts between hills like a sandy ribbon. Beyond, the valley appears brown, even the Sierra has lost its chalky blond hue. The deep green of trees marking waterways and towns is washed with burnt umber, as though the brown land of the Inyo casts a muted dust into the air. The land is seen through a lens of chocolatey dryness. A warm wind shakes the few remaining dry tufts of grass that cattle and deer overlooked. Most bunch grass has been otherwise gnawed to blackened stubble. To the east, the Inyo Mountains rise multicolored toward brewing afternoon thunderclouds. Their high slopes and peaks blackened by cloud shadows and pine forests.

Although the lowlands are most changed by increased emigrant population, the pinyon pine forests in the White-Inyo Range were heavily harvested as fuel wood when mining boomed, especially after silver was discovered at Cerro Gordo by Pablo Flores in the late nineteenth century. Prior to nonindigenous settlement, pinyon-juniper woodlands were important to the Numu, who harvested the protein-rich pinyon pinenuts. The forests still darken the higher elevations like stubble on a morning-after face, though they've long since lost their connection to Numu well-being.

In the high reaches of the White Mountains (10,000 to 11,650 feet) grow the ancient bristlecone pines, some of which are the oldest living things in the world. Existing specimens are over 4,700 years old, sprouting from seed when the first of the great pyramids of Egypt were under construction. It has been estimated that the forests were here as far back as 8,200 years ago, in interesting coincidence with the earliest known human occupation of the Owens Valley. Perhaps people and trees could inhabit land newly freed from the ice as the glaciers of the Pleistocene Ice Age retreated. While other plants have waxed and waned in response to changing conditions, the bristlecones have managed to survive. People who live in harsh places are often better able to adapt to environmental changes than are people pampered by soft living. So it seems with plants. The oldest bristlecones grow in the harshest portion of their range, where chalky dolomite soil is dry and nutrient-poor and winds blow cold and constant. Their slow growth rate, less than one inch in diameter every century, helps the wood resist rot and bugs. The saying "live fast, die young" suits the bristlecone: the trees in the more benign areas grow more quickly but succumb to disease more easily. Drawn by the bristlecone's resilience and age, thousands of visitors make the winding drive up Westgard Pass Road just to walk among the old trees.

SEPTEMBER 26. Schulman Grove, the Ancient Bristlecone Pine Forest. I meet a cool, biting chill as I walk the Discovery Trail through the pines, the trees that helped Dr. Edmund Schulman realize the age of the bristlecones. The trees grow on rocky blond dolomite slopes and look like the foxtail pines of the Sierra Range. The oldest ones are stunted and fat with bleached bark, more the color of the chalky soil at their roots than the cinnamon bark of the foxtails. My fifteen-month-old daughter, Isabel, is tucked behind me in a backpack, refusing to wear her fleece hat. Being with her among this enduring forest helps me to feel time and history in the land and trees. Like leaving a new friend, she says "bye bye" to each pine we pass along the trail. When the trail winds out onto an exposed slope the wind becomes fiercely cold, so I jog the remaining half-mile to the parking lot. Isabel learns to say "running" and "freezing" by the time we reach the truck. It starts to snow and we drive back down the road, stopping just below the bristlecones to gaze across the Owens Valley. I think of all that these trees have borne witness to, all the human wanderings, fighting, living, and dying down below in the valley. Isabel, her cheeks chapped rosy from the hike, sleeps while I ponder the over 135 generations of people who have been born, lived, and died in the land below while the bristlecones have yet to complete one life cycle.

This pendular sweep from west to east across the valley reveals the landscape's tremendous plant diversity. In a scant twenty miles the landscape supports alpine blooms to desert scrub and back again, making the valley a diverse and rich place for wildlife and people. The mountain ranges casting long shadows onto the narrow band between them, the Owens River sucked dry and left for dead on the valley floor, and the pines growing darkly on the mountain sides—these are the ingredients of the Owens Valley landscape. Despite their desiccated appearance, taken together these three elements—mountains, water, and vegetation—create a landscape ripe for human occupation.

58 *This is the sense of the desert hills, that there is room enough and time enough.*

MARY AUSTIN, *LAND OF LITTLE RAIN*

Eastern Sierra Range, with Mount Whitney in the distance (center).

Eastern Sierra Range, looking east on the Whitney Trail.

The lesson is that the whole thing—the whole Basin and Range, or most of it—is alive. The earth is moving. The faults are moving. There are hot springs all over the province. There are young volcanic rocks. Fault scars everywhere. The world is splitting open and coming apart.
KENNETH DEFFEYES, IN JOHN MCPHEE'S *BASIN AND RANGE*

Alabama Hills (*above*); Inyo foothills (*below*).

Looking west across the bottomlands to the Sierra Nevada.

Dawn at Ramshaw Meadows on the Kern Plateau, Sierra Nevada.

I immediately saddled my horse and rode down to the river, and finding it almost impossible to talk from bank to bank in consequence of the sloughs on either side of the river being swimming, I resolved to cross myself. After swimming two sloughs and the river and wading half a mile through willows and tulles, I reached the eastern bank of the Owen's River.

LIEUTENANT COLONEL GEORGE EVANS TO MAJOR RICHARD C. DRUM, JULY 9, 1862

Owens River, ca. 1914 .

Owens River, 1999.

This lake is of the color of coffee, has no outlet, and is a nearly saturated solution of salt and alkali.

WILLIAM H. BREWER, *UP AND DOWN CALIFORNIA IN 1860–1864*

Owens Lake, ca. 1914 (*above*) and 1997 (*below*).

Foxtail pine, Sierra Nevada and Bristlecone pine, White Mountains.

Cottonwood tree, a remnant of the rural landscape, Owens Valley.

Inyo Mountains.

2 DWELLING BEFORE

AUGUST 4. Owens Valley from 13,000 feet. In the view to Owens Valley from a ledge near Tunnabora Peak of the Sierra lie river, fault, trail, rail, roadway, fence, power lines, phone lines, and aqueduct—the linear and thin sinuosity that define the valley's structure. Within this virgate landscape, braids of lines weave intricate patterns through the bottomlands and down the alluvial belt that skirts the Sierra Range. Writhing like mated snakes, nature and culture twist around each other becoming inseparable, indistinguishable. Nature and culture melded in the muted landscape, dusted brown, streaked with green. "Nature abhors the line," claimed nineteenth-century landscape garden style enthusiasts, but the ice crystals shifting on the glacial lake that sits in the shadow of Tunnabora and the fractured granite heaped in rectangular blocks at this breathless height seem to rejoice in angularity and straightness. From so high a perspective, sitting on a granite shard perched 9,000 feet above the valley, I can read the landscape's geometry and know that nature adores lines. The word is linked by its roots to the fiber of flax, the plant used to make linen. Nature celebrates the lineated simplicity of shadows cast across the valley floor at sunset, fractured ice that floats on glacial lakes, and sheer breaks in the granite that create the Sierra's eastern range.

This is the view from Tunnabora Peak, but from the valley floor the linear structure, so obvious from my alpine perch, vanishes in the distraction of cultural details. Stretched along U.S. 395 the small towns of Olancha, Cartago, Lone Pine, Independence, and Big Pine, each aligned to a Sierra peak, are shrunken oases along a thread of roadway. Punctuated by tall cottonwoods, Lombardy poplars, and Italian cypress, they read as quick accents between the road's long paved pauses. During winter weekends, a constant stream of ski-racked four-by-fours on their way to slopes up north flood the narrow strip of highway. A quick stop at a corner coffee shop and they're off again. The movement creates a hypnotic pulse, the drone of snow tires throbing like blood through a narrow artery. Economic sustenance pumps through these small towns, drop by drop into cafes, motels, and service stations, keeping the main street from drifting off into an eternal deep sleep. Behind the highway are a few blocks of widely spaced small homes and trailers. Dogs ride in the back of pickups, tongues wagging, their excitement seeming out of place in the drowsy streets. Wallace Stegner has described towns such as these: "They look at once lost and self-sufficient, scruffy and indispensable. A road leads in out of wide emptiness, threads a fringe of service stations, taverns, and a motel or two, widens to a couple of blocks of commercial buildings, some still false-fronted, with glimpses of side streets and green lawns, narrows to another strip of automotive roadside, and disappears into more wide emptiness."[1]

In Lone Pine, the economic base is mostly tourism, Department of Water and Power (DWP) jobs, and some cattle ranches that are "just scraping by," says a woman at the Chamber of Commerce office. Selling scenery, outdoor recreation, and nostalgic forays into the West to urban tourists is the valley's strongest economic resource. The generations reared on television's *Little House on the Prairie*, *Big Valley*, *Bonanza*, *The Lone Ranger*, and *The Wild, Wild West* take pleasure in weekend escapes to this remote valley to spend time breaking in brand-new Levis in a hired saddle, loping along some lonesome trail or duding for a day on a cattle drive. The landscape provides a dramatic backdrop for urbanites to act out their cowboy fantasies. The towns themselves maintain vestiges of bucolic imagery, but only remnants linger of the rich farmland that existed at the turn of the twentieth century. No longer fed by surrounding fields, these towns have turned their faces towards the highway, letting their agricultural link to the land slowly fade away. A full day's ride on horseback from Independence to Lone Pine has shrunk to a twenty-minute drive in a car. Highway 395 cuts through the isolated rural villages, transforming the settlements into linear towns.

Or so it seems from the road. The towns appear to exist for the road, because of the road—the intravenous drip that feeds each small community. But a closer look at the highway when it becomes Main Street in Lone Pine reveals evidence that these small towns support more than tourism. In Lone Pine, whose welcome sign reads "Little town. Lots of Charm," one can find almost all the ingredients for American living on Main Street. I measure a town's sincerity and vitality on the shelves of its hardware store; a town isn't a town if you can't find a good hammer in it. By this gauge, Lone Pine's single strip of Highway 395 is alive and well. From airport to cemetery, there is every kind of roadside attraction—town parks at both ends, two hardware stores, cafes with red vinyl booths and pictures of the Duke on the walls, a video store, sporting goods stores decorated with neon trout signs and hand-drawn AMMO advertisements, Joseph's Bi-rite Market, a barber shop with showers, Jake's Saloon, a Mexican restaurant, motels on the ends of the strip, a thrift shop, plant nursery, gas station, drug store selling sundries, donut shop, two Western clothing shops, a photo processing store, vacuum and appliance shop, electronics and leather shop, auto parts store, bank, bike store, camping equipment shop, hair salon, real estate office (selling the same houses over and over again because the DWP owns most of the property), a bar named after a cattle brand—the nefarious kind that exudes a simultaneously alluring and repulsive pungency of cigarettes and spilt beer, coffee shops, a Greyhound bus flag stop, pizza parlor, the Chamber of Commerce, an art gallery, liquor store, trading posts, gift and craft shops, a high school, baseball field, frosty shop, DWP and National Forest Service offices, a minimart, towing company, Feed and Seed store, propane gas supply store, golf course, lumber yard, mobile home park, cafe / laundromat, fuelwood lot, ministorage facility, and an English Pedunculate oak that came from Sherwood Forest around 1900. Everything one might need except a movie theater or whorehouse, all with a Lone Pine address on U.S. 395. This cluster of goods and services separates the valley's towns from the faux historic hamlets that are cropping up all across America, the ones stuffed with quaint shops selling old American bric-a-brac, Amish quilts, and expensive fudge. Towns in the Owens Valley are genuine small Western towns with good hammers for sale.

Today these towns, along with Bishop to the north and adjacent reservation lands, are the most peopled places in the Owens Valley. They are the valley's dwelling places, yet they tell only part of the story of living beside the long thin lines. The Owens Valley has been a place of human occupation since at least 8,000 years ago, long before the river last spilled over the Haiwee Divide and into the Mojave Desert as the climate shifted toward aridity. Landscape changes, though significant on a geologic time scale, have been gradual, as have been, for a long time, the subsequent changes in human settlement and land-use patterns. The slowly shifting human occupation over eight millennia are part of the cultural landscape of the Owens Valley. Since the mid-nineteenth century, human settlement has paralleled more cataclysmic geologic patterns, going along in one direction until some social fault shifts or some human volcano erupts, causing the course of life to meander, to go another way.

Beginning with the intrusion of miners and ranchers during California's drought of 1859, life for valley dwellers has been a series of upheavals. Anyone living in the valley from this time on had their life jolted from its foundation, displaced. Numu culture was all but destroyed by American occupation; mining culture by its nature was ephemeral and susceptible to sudden decline, so its upheaval when a claim ran its course was perhaps the least surprising; pioneer settlers lost their livelihood and way of life when many sold their water rights to the city of Los Angeles; and Japanese Americans interned at Manzanar War Relocation Center during World War II were thrust into a strange land during a time of fear and war and could not have been more culturally shaken. Yet between tremors, people were born, grew up, had families, and found some connection to the landscape. Whether native or newcomer, or interned at Manzanar, valley dwellers describe this land between with love and passion.

People are inextricably tied to the land, to a place, through birth, memory, nourishment, and love. How people move through, settle in, mark, or use a landscape defines this connection. The stories of native peoples, miners, settlers, and Manzanar internees, exhumed through layers of time, come alive in the land they touched, the people they loved. The intimate, messy day-to-day of living and dying makes history breathe; the rest, the facts (if there are any), are stuff of legends and useless generalities. *American settlement and conquest of the indigenous peoples was inevitable. Los Angeles stole the Owens River water. The Japanese at Manzanar were a threat to national security. John Wayne was the greatest American cowboy.* While these historic one-liners can effect cultural change, they do little to clarify the complex and dynamic human relationships to landscape. If anything, they numb memory by reducing history to simplistic sequences of general events.

Human history is more than a compilation of facts strung together in a linear chronology. It is a collection of individual stories, reminiscences that contradict one another, leave countless loose ends, yet convey the complex and vivid nature of humanity. Human history is made of tales that begin with "I remember when . . ." As a way to get closer to the experiences of valley dwellers, I wrote short fictional pieces inspired from reading many such recollections. The three narrators in these vignettes —a Numu mother, a mining laborer, and a ranching woman— simply distill moments that are meant to remind me and readers that the three generalized groups of people discussed in this essay—Numu, miner, and emigrant—were made up of living, breathing human beings. People who gave birth, missed their wives, and baked bread.

I am *via,* mother: Labor began with gradual urgency. Early on I went about my morning chores, mending the winnowing basket I crafted from willow last winter, going with the other women to collect *tupusi.* It was my first child and would be a long time coming so I turned my attention to other things. They had been preparing the *cutuvida* [birthing pit] since the labor began and would soon begin to warm it with fire. I had done all the right preparations, staying away from many meats so the baby would not be too big and cause pain in the delivery. I knew the stories of birth from my mother and grand-mothers, but there was still surprise as the contractions mounted.

The pit smelled fresh of fire, burning sagebrush from the desert, and of soil. My joints felt slack from steam drifting off the hot rocks bubbling in a water basket. The *tuduamudukudu* [midwife] lowered into the pit behind me. I felt surrounded, enveloped by earth and sky, cradled by the *tuduamudukudu*. The air thickened with sweat and the musty odors of birth, almost suffocating my calm. Pushing, my body wracked by escalating waves of contractions, weakened by the long labor, I wondered if something was wrong. Then the *tudu-amudukudu* began to instruct me in earnest, grasping me from be-hind, massaging the baby down. I let go one stifled cry, a mixture of pain and relief. The cries from behind me and the emptiness in my womb told me the baby had come. *Waisua,* tiny baby girl, was bathed and brought to my breast.

Numu myths and legends recall a time when the land sported a cooler, moist climate, when Jeffrey pines reached down towards the valley floor and the lake spread large like a blue jewel. The val-ley has changed greatly since this ancient time; the lake has dried and forests have shrunk away from the hot valley floor. Although this climatic trend toward aridity began long ago, the Numu still enjoyed more surfacewater and groundwater in 1850 than inhab-itants of the valley do today, with "extensive marshes and grass lands watered by streams from the snow-capped Sierra Nevada mountains."[2] As their population grew, they subsisted on scarcer food supplies in a dry land but still had ample water for a people dwelling in the desert. Local historian Charles Campbell notes, "Of the many wandering bands of people in the Great Basin, life was probably best for the Paiute people in the well-watered Owens Valley."[3] The valley supported the densest population in the Great Basin region, with 2,000 inhabitants, averaging one person per square mile—still less than a tenth of the valley population's current density.

The native peoples, like the settlers who came after them, were drawn to the water's edge. Water penetrated much of their interactions with the land. The name originally attached to these inhabitants by scholars is derived from water: "The Pah-Utes are Water Utes, taking their name from their rarest and most precious resource."[4] Anthropologist Julian Steward found a man from Fish Springs who used the phrase *nungwa paya hupe caa otumu* (we are water ditch coyote children) to name the people of Owens Valley.[5] Numu creation myths describe them as coyote's children, and the Owens Valley was called the water ditch. Taken in its context, the valley was indeed a water ditch, as in 1850 it had one of the few perennial streams in the Great Basin Desert. Instream flows and groundwater were ample, but, with five to six inches of rain per year, the valley was still dominantly desert scrub, with lush vegetation only along water courses and lowland meadows. Except during pinenut harvests, Numu camps were located in these well-watered areas. The Numu called the valley *Payahu Nadu*, a name derived from their word for water, *paya*.

In early fall, while the pinyon cones were still green, the Numu moved up into the White-Inyo Range to set up pinenut camps. If the harvest promised to be a good one, *wogani* were built of pine or juniper limbs draped with pine boughs. Outside the entries to these simple gabled structures, two pits were dug, one for storage and one for roasting. The cones were harvested green to avoid competition from animals who also relied on the pinyon for food. They were then sun dried or heated to release the nuts, called *tuvay*. The pinyon pine woodlands occur from a 6,500- to 8,500-feet elevation on the dry slopes. Winter snowfall would edge downhill through the season, reaching coldly into the pinenut camps. The camps were located in proximity to the pinyon and

were not necessarily near a stream or spring, so winter camps relied on snow for water. There are remains of small rock dams in
shaded ravines near old pinenut camps; these dams are thought to
have been used to store snow for a water supply. This harvest practice required tremendous effort, revealing the importance of the
pinyon as a food source and demonstrating the Numu's devotion
to gathering food. After the pinenut camps were abandoned for
the season, the Numu would move back down to the valley.

As a people, the Numu, in their migratory patterns, followed
the logic of the land. Vegetation comprised 70 percent of their
food source, so settlement and movement echoed harvest cycles.
The women collected seeds and the tuberous roots of plants such
as *mono, nahavita, posida, tupusi,* and *wai*—thought to be lovegrass,
blue dicks, tomcat clover, yellow nut-grass, and Indian mountain
grass, respectively. The migration route could change from year to
year in response to climatic variations that caused fluctuations in
seed or tuber ripening. When camps were established, domed *toni
nobe* were built for shelter from willow and tule reeds. The willow
formed a lattice frame, which was then thatched with tule bun-
dles. These structures, used mostly for sleeping and storage, were
only ten to fifteen feet in diameter. The single door opened to the
east. A wind screen was also constructed near the *toni nobe* to protect the outdoor cooking area. Both the *wogani* and *toni nobe*
reflect the seminomadic nature of the Numu. Their structures,
whether in the mountains or along the valley floor, were simple
and could be quickly constructed from nearby materials. Weather
damage incurred while families were away collecting food could
be easily repaired upon their return.

Julian Steward, who conducted ethnographic research in the
valley from 1927 to 1934, presented the Numu as a simple people
who expended little time and energy toward anything other than
sustenance. Ceremonial practices were straightforward and sparse.

They gained spiritual strength from the natural environment, usually through contact in dreams with animals or places. "Individuals' powers embrace most things in nature. Eagle, fox, bat, snow, obsidian, the blue haze sometimes over the valley, and Birch mountain and Mount Dana in the Sierra Nevada have been powers."[6] Powers reflected an outstanding quality of an animal, natural feature, or phenomenon. Eagles granted speed, mist granted protection, and a mountain could give strength. Identifying personal power to an element in the landscape reveals the Numu's strong sense of connectedness to their surroundings. Physical configurations of places in the valley also seem to carry importance, as can be seen in Numu place names, many of which reference outstanding physical features of the area. The following list, adapted from Steward's article, "Ethnography of the Owens Valley Paiute," gives translations for place names:

> *Tanova witu* (a village near current Manzanar site)
>> *tanova* = saltbush
>> *witu* = place
> *Tupusi witu* (a village at current Manzanar site)
>> *tupusi* = yellow nut grass
> *Tumutsadu witu* (place east of Alabama Hills across the
>> Owens River)
>> *mutsa* = the rounded end of a hill
> *Tsiguhumatu* (a creek east of the Alabama Hills across
>> Owens River)
>> *tsigopi* = rabbitbrush
>> *hu* = creek
> *Wakopo witu* (Lone Pine Creek, where a large pine
>> once stood)
>> *wako* = pine
>> *po* = alone

Paokarangwa (Birch Mountain)
  pao = rocky
  karangwa = peak or boulder peak[7]

Another illustration of how the Numu interacted with the land comes from two interviews Steward conducted in 1927 and 1928. The long excerpts from both interviews that follow bring light to how the Numu interpreted their surroundings and events in their lives. Both men were nearly 100 years old when they were interviewed.

Jack Stewart, Hoavadunuki, enjoyed successes throughout his long life within the Numu community. He lived most of his life in the village territory of Tovowahamatu. Territories included the settlement sites where the Numu stayed most of the year; Tovowahamatu was around the present town of Big Pine and nearby hunting and food gathering lands. Hoavadunuki attributes his respect and stature within the community, his power, to a mountain in the Sierra Nevada Range, a 13,602-foot peak visible from his home territory. "When I was still a young man, I saw Birch Mountain in a dream. It said to me: 'You will always be well and strong. Nothing can hurt you and you will live to an old age.' After this Birch Mountain came and spoke to me whenever I was in trouble and told me that I would be all right. That is why nothing has happened to me and why I am so old now." Throughout his life his mountain gave him strength and aided him in hunting. For the Numu, being a good hunter was a measure of masculinity. Before going into the mountains to hunt large game, Hoavadunuki would call upon his power: "Now, great mountain, I wish that you would give me some of your deer to eat. You have so many on you. If you would give me some, I wish you would have them at your foot, not far up." After his power had helped him kill a deer and a mountain sheep, he attributed his own ability to carry the load down the mountain and across the valley to his power: "The deer and mountain sheep were a heavy load, for I had packed them both at once down to the valley. But when I was a young man nothing was too heavy for me. I enjoyed carrying a large, heavy load. Didn't my power come from the mountain upon whose back are rocks which never hurt it?"[8]

Hoavadunuki traveled extensively within the Owens Valley region, only once leaving the area to live near San Francisco with a friend. After a few years away from home, when he was beginning to tire of living with his companion, his mountain called him back to his land and people:

> I saw my mountain in a dream. It rose up in the east and looked for me. It looked first to the south and then to the west, and when it saw me near San Francisco it said: "You must come back soon to your own country and your own people. Nothing will happen to us. We will always be just where we are. Don't forget to come back. When you get back, you will kill a large deer—the largest yet" . . . Few Indians leave their own country who do not return . . . When I got home, I went hunting and killed the largest deer I had ever got. This happened just as my mountain had promised.[9]

From Hoavadunuki's viewpoint, success, stature, community respect, and solidarity were granted to him by his power, Birch Mountain.

Sam Newland, whose Numu name was not recorded, led a different life. Unlike Hoavadunuki, he never dreamed of a power, which, in part, may explain his poor performance in hunting, dancing, gaming, and gambling—all important measures of masculinity for the Numu of that time. "When I was a young boy I knew nothing about hunting, so I did not try it often. I guess the reason was that my father died when I was very young and neither my uncle nor other relatives took the responsibility of teaching me. I never did dream of a power. I played games with the boys sometimes but had very little success. When we played throwing arrows at a hoop, I generally lost all my arrows; for in the first place I did not know how to make good arrows and in the second place I could not throw them." Although told with less bravado than Hoavadunuki's tale, his story reveals some aspects of Numu landscape perception. Sam Newland spent much of his time traveling throughout the Owens Valley region, usually going where food could be provided for him since he was such a poor hunter. "We boys decided to take a trip east, so we started out and went to a pinenut camp called tupiko. The next day we went to another pinenut camp, hunaduduga, where we spent the night. The people gave us pinenut mush . . . We went north up through the White mountains and visited several more pinenut camps . . . We went to the 'fandango' and . . . were given supper soon after arriving and then I went and watched them dance. The men and women were doing the circle dance, but I did not join in; I was a poor dancer."[10]

Hoavadunuki and Newland are very different in self-image, but their stories carry some common threads which reflect their culture's perception of the land. Much of their life stories involve local travel and acquiring food. In a landscape edging towards aridity, gathering food demanded more and more attention, so it is no wonder the task is prevalent in their memories. The Numu lived lightly on the land, both through their migratory practices and by utilizing local vegetation and wildlife. By 1800, due to climatic changes, timber forests had disappeared from the lower elevations and only alkaline soil remained along the valley floor; their land offered little for them to exploit. Unlike the native peoples of the Midwestern Plains, who relied on bison as a primary

86

food source, the Numu depended on many sources for food and shelter, which wove them deeply into the natural web of life. They kept their numbers small and lived, in the language of today, a self-sustaining existence through the practice of sensitive land ethics.

So, until the nineteenth century, human beings had only subtle influence on ecological and physical transformations taking place in the Owens Valley. The Numu's impact on the land was made by way of subtle methods used to maintain or enhance stable food sources. Food collecting, particularly of pinenuts and fly larvae, did affect the populations of the collected species, their habitat, and the populations of other species that fed on them. For example, farther north near Mammoth Lakes, the Numu once harvested the Pandora moth caterpillars from the Jeffrey pines. When collecting diminished due to conflicts with settlers and the government, defoliation began—because these larvae feed on pine needles. Stream diversion to irrigate fields and trap fish and brush burning while hunting deer are two other practices that had an impact on the Owens Valley ecology during Numu occupation. An irrigation system consisted of damming the water source, generally a creek off the Sierra Nevada Range, and creating a series of ditches to each plot to be irrigated. In the springtime the irrigator, *tuvaiju* (from *tuvayadute*, meaning "to irrigate"), determined when and how much water would flood a field, which would later be harvested by women for seeds and bulbs. The position of *tuvaiju* was one of high honor and responsibility, indicating the importance of irrigation in Numu communities. Fish were gathered in the dry creek bed after diversion, and then the cycle reversed at harvest time; the dam was dismantled, water resumed its natural course, and fish were gathered from the ditch. *Tuungwava*, a plant in the lily family also called slim solomon, was crushed, wrapped in old baskets, and placed in the water to stupefy the fish and make catching them easier.

The Numu had extensive irrigated fields throughout the valley which they may have planted with seed or tubers. In these fields, seeds were collected from spring to midsummer and roots were gathered in late summer. The irrigation technology was sophisticated in light of their rudimentary tools. A news correspondent accompanying Captain Davidson's expedition in 1859 praised the refined engineering of their irrigation system in a letter to the public: "Large tracts of land are here irrigated by the natives to secure the growth of the grass seed and grass nuts—a small tuberous root of fine taste and nutritious qualities, which grows here in abundance. Their ditches for irrigation are in some cases carried for miles displaying as much accuracy and judgment as if laid out by an engineer, and distributing the water with great regularity over their grounds."[11]

Large game, such as deer, helped to supplement the Numu diet. Each village territory included hunting grounds in the Sierra. Hunting was done individually or communally, the latter method using fire to enclose deer. Men advanced with sage bark torches, setting brush on fire as they went. Deer were trapped in a circle of fire and shot. Women and old men stayed in camp and cured the meat. These kinds of human interventions have been occurring in the Owens Valley and surrounding mountains for thousands of years, so pristine nature, unaffected by human beings, existed here only before the landscape broke free of glacial ice and began its amble towards aridity. For at least 8,000 years, the Numu and indigenous peoples who came before them touched and changed the landscape.

Yet the seemingly soft and subtle changes effected by the Numu were crudely trampled by American progress. The Numu knew how to live in balance with what the valley could bear; when they looked at the land they saw precisely what it was and not what it could become. They saw a potent land that needed to be treated with humble respect because it had powers far greater than those of people; a land that could provide for them as long as they took only what they needed. In the land they saw the source of their strength and felt their own fragility.

In contrast, the Owens Valley that the early American emigrants and military men found was a remote Eden, an untapped oasis. From a nineteenth-century American perspective, the valley was a land of opportunity. Captain Davidson, who came from Fort Tejon to investigate allegations that native peoples of the Owens Valley were stealing stock from the Los Angeles area, wrote a glowing report of what he found on his 1859 expedition:

> Wherever water touches [the soil], it produces abundantly. I should think it well suited to the growth of weath, barly, oats, rye, and various fruits, the apple, pear, &cc. The grasses were of luxuriant growth. In one meadow . . . the grass (over two feet in height, broad leaved, and juicy) extended for miles . . . To the Grazier, this is one of the finest parts of the state; to the Farmer, it offers every advantage but a market . . . There is building timber enough for all the uses of a population commensurate with the agricultural resources of the valley . . . it is the finest watered portion of the lower half of the state.[12]

It's doubtful that the Numu saw the land in this light. Their desire to keep in balance with what the valley could offer was often viewed as unambitious and un-American by the settlers who eventually conquered the native peoples and took their land. When miners began to work ore in the mountains and cattlemen drove their herds onto the lowland meadows, the Numu's delicate balance between sustenance and starvation started to collapse—their precious pinyon became fuel wood for smelting and seed and bulb fields were trodden by cattle.

Perhaps if the Owens Valley were not so remote, there may have been more outside interference before the early mining operations and cattle drives. The valley did not lie along the routes of either the Spanish missionaries or the majority of American emigrants, and, prior to the mine strikes east of the Sierra, prospects in the Inyo Mountains were outshined by claims further west. Before 1859, the Numu had had only limited interaction with a few pocket hunters—lone prospectors searching for shallow ore-rich crevices in the earth—and early explorers. The Spanish missionaries didn't venture into what they thought was an unpromising inland valley, though the Numu probably knew of them through trade with indigenous peoples to the west.

The Numu exchanged goods such as pinenuts, red paint, and sinew-backed bows with the Monache, Tübatulabal, and Miwok in the Sierra. When Captain Davidson traveled through the valley, his interpreter tried, with limited success, to communicate with the Numu in "horrible Spanish."[13] The Numu may have learned Spanish from tribes to the west or from the early expeditions that had passed through the valley. In the early nineteenth century, the military and fur-trapping groups that traversed the valley were bound for or returning from Mexican-occupied California. In 1821, Mexico won independence from Spain and began governing this remote province of her newly formed country. Walker's 1833–34 expedition had wintered in the Mexican province, so men no doubt spoke some Spanish by the time they returned to the States by way of Owens Valley. Zenas Leonard, the diarist of the trip, notes numerous encounters with "Spaniards," beginning with his November 23, 1833, entry. Around March 2, the party entered a village in the Sierra foothills where inhabitants had "eight or ten years since resided in the Spanish settlements at the missionary station near St. Barbara." They "found that these people could talk the Spanish language," and, thinking this advantageous for their group, they hired two as pilots to accompany them across the Sierra.[14] The people Leonard met had probably fled La Purisma Mission (in present-day Lompoc) during the 1824 revolt. On March 15, 1834, Leonard writes of meeting two men that had deserted from the Spanish army. It appears that the men joined the party to avoid military punishment for desertion. It's not clear whether or not they remained with the party when it crossed over to the Owens Valley.

The Numu were not uninformed about people and events outside their territory, but until 1859 their land and food sources had not yet been directly threatened by outside contact. In spring of that year, gold was discovered in Mono Basin, north of Owens Valley, an event that forever changed the people and land use in the valley. Monoville sprung up around the diggings, a short-lived boomtown that quieted by 1862. In June 1859, the silver deposit known as the Comstock Lode was discovered in the Washoe District at Virginia City, Nevada. More strikes followed in the Esmeralda (renamed Aurora) District in the fall of 1860. These mineral discoveries, all north of Owens Valley, drew thousands of miners over the Sierra, from Nevada, Mexico, and around the globe. By 1863, Aurora hosted a population of 5,000.

During this time, the Owens Valley became a "great thoroughfare," as Lieutenant Colonel George Evans described it in April 1862, the best route "through which the growing trade and travel of this southern country must pass in and to the Esmeralda and Washoe districts."[15] The promise of gold and silver lured men to these hot and rocky regions where, like swarming locusts, they followed strikes. Most men left their families elsewhere while they chased wealth.

I am a miner, digging: It's back-breaking labor they say and they's right. Hunkered over my pick ax, digging and hacking away at the confounded earth, hands so worked over I can barely recognize them as my own. The self-same hands that used to stroke the soft cheeks of my young bride. Now here they are hanging off the ends of arms that haven't wrapped around that fine woman in so long they're forgetten how. I'll be coming right home, I says to her, just as soon as I make enough to buy ourselves some good farm land in Visalia. Cerro Gordo was booming and the mines were paying good wages, so I set out along the Hockett Trail through the mountains. It's not but two or three days hard riding across the Sierra Nevada so you needn't worry I tells her. I'll be home before you've learned to miss me. But it's a strange country here in the Inyo Mountains across the Owens Valley from the snow-covered Sierra Range, a hard dry-scrabble wicked place that keeps you even when you're beat down and aching to be home smelling the dampness of fresh-cut alfalfa. I set out to leave each year before the snow stops up Olancha Pass. I got as far as the crossing on the Kern River once, but I got to thinking about the mines, the hard dry mountains, and the digging. My horse, almost reading my thinking, stopped short of the water's edge waiting for me to tell it what it already knew . . . we're turning back. There's no romance in working the mines. It's just plain hard and dangerous work. And living in that town, not really a town so much as a scramble of folks just passing through and needing a place to hold up a while. Some, like me, have been fixing to leave any day now for five or six years. So no one owns more than they can carry on a mule or strap to their saddle. No sense bringing in all the civilized necessities like we have in our farm town. There's barely enough women around to keep a town like this civilized and them that are here, well. So here I stay, waiting for this land to let loose of me. I've bought that farm ten times by now with the money I've sent back to my wife. She's stopped asking when I'll stop digging. Soon, I says, any day now.

A LAND BETWEEN

In 1865, when the Aurora diggings were played out, silver was discovered at Cerro Gordo in the Inyo Mountains. When miners, cattle, and settlers moved into the valley, the landscape's choreography shifted as the Numu circle dance was replaced by the line dance of free trade capitalism. Numu migration from food source to food source was supplanted by the movement of goods and capital to settlements. This subtle but significant change from migration to transportation destroyed the Numu way of life and altered the valley and mountain landscape more in ten years than people had done in eight millennia. The simple cyclical transformation of seed, nut, fruit, fish, or deer into food was dismantled and replaced by the complex linear abstraction of converting rock into commerce.

93

Turning a stone into a loaf of bread demands tremendous investment in labor and materials. The ore must first be extracted from the mountain. This takes people who need to eat, drink water, and have shelter. Farms and ranches cropping up in the west valley supplied food; water was packed in by mule; shelter was provided in the hastily built town of Cerro Gordo. The ore must then be smelted. This takes fuel, and in Cerro Gordo, this fuel was in the form of trees turned to charcoal. To the horror of the Numu, the nearest timber came from the pinyon-juniper woodlands of the White-Inyo Range. After these fuel sources were exhausted, trees were felled in the Sierra, hauled by oxen teams, sent down the mountain in wooden flumes four miles long, turned to charcoal in kilns by the lakeside, then sent across the lake by steamboat and hauled to the smelters. Once smelted, the bullion needs a market, a city with enough commerce to give value to the silver. Los Angeles provided that market, and the bullion was hauled there by mule train, as much as eighteen tons a day making the 440-mile trip across the Mojave Desert in twenty-one days. In Los Angeles the bullion was traded for cash, with which a baker could purchase supplies to make bread and someone else could buy a loaf. It's a long string of arduous tasks before a person is fed. Meanwhile, the pinyon pines are depleted and a culture starves. This great effort drew miners, ranchers, and farmers, all participants in the line dance, into the valley and began the tumultuous period of landscape transformation and cultural upheavals.

Accounts of the clash between the Numu peoples and newcomers to the valley often present it as a battle between natives and pioneer Americans, but this simplistic interpretation ignores the role of the mining industry and its diverse population. Because silver was a predominant ore extracted from the mountains east of the Sierra, the strikes attracted miners from Mexico, who were renowned for their expertise in silver mining. Many discoveries, including Cerro Gordo, were made by Mexican prospectors who had learned the silver-mining trade in Sonora, the Mexican state bordering California and Arizona. They used *arrastres*, cylindrical tanks made of stone, to process ore. Draft animals turned the milling stones to crush the ore, which, mixed with water, became a paste. Gold or silver was then precipitated out with quicksilver (mercury). This method was cheap and self-sufficient but slow, grinding only one ton of ore per day. *Vasos*, ovens for smelting, were also used in the early mining days.

By the 1870 census, two-thirds of the region's miners were either not American citizens or were foreign-born citizens—about half were from Mexico and most of the remainder came from Ireland, England, France, or Germany. Miners from the United States came predominantly from coal mining regions of Pennsylvania, Illinois, and Kentucky. Cerro Gordo also had an established Chinatown along the road to the cemetery. The China Stope, a chimney of Cerro Gordo's workings, was named for the Chinese men left buried under 500 feet of rock when other miners blew it up. This "was a union thing," explained the current owner, who showed me around the Cerro Gordo ghost town. Cerro Gordo's remote mountain location, accessed by the steep Yellow Grade, so named for the ochre rock that the road cuts through, lent itself to the town's lawless reputation; legend says there was "a murder a day."[16] The diverse collection of people, separated by language and racial enmity, often had only their desire for gold and silver in common. The emigrants of the valley bottomlands suffered similar disunity, but they seemed better able to develop lasting settlements. Through the efforts of this eclectic collection of newcomers, the land was transformed.

Mining and agricultural settlements developed different characters and structure. In the Owens Valley, though, they maintained a symbiotic relationship from opposite sides of the valley, each town tucked against its own range so the inhabitants could extract the mountain resources. To the east the White-Inyo were gutted to release gold and silver ore, dolomite, zinc, and limestone. To the west the Sierra Range released water from its snowy peaks, sending it tumbling down its canyons along with fluvial deposits. Miners extracted, farmers cultivated and irrigated, and both exchanged their goods across the valley by rail and mule team. Despite this symbiosis, mining and farming settlement patterns and land impacts differed beyond their geographical locations. Although the towns shared gridiron plat plans in their early conception, the actual organic accretion of structures, civic identity, community, and even graveyards of one bore no resemblance to those of the other. East and west reflected little similarity from across the valley, yet they were codependent. Each side catalyzed the establishment of the other, though the mining towns didn't have the same capacity to survive as the farm towns.

SEPTEMBER 23. Driving the Yellow Grade from Cerro Gordo. I'm trying to imagine the fourteen-mule teams loaded down with nearly eight tons of silver bullion making their way down this road. The edge drops over 4,000 feet away to the lake below at a sickeningly steep slope. Mottled with cloud shadows, a wash of muted colors, the lake surface seems to swim with light, as though the salt-encrusted playa has turned to liquid. At 7,000-feet elevation Joshua tree, a plant I identify more to the Mojave Desert, begin to dot the yellow rocky landscape. Isabel, so unaffected by the treacherous descent, has fallen asleep, forcing me to swallow my inclinophobic impulse to stop the truck and crawl on my belly back to the safety of my cozy flat bed at the Belshaw House in Cerro Gordo. I could settle there, anything to keep from driving down the narrow road paved only with tractionless shale.

Abandoned mining towns are usually called ghost towns, because they still harbor a ghost of hope that another boom is close at hand. Cerro Gordo's owners say there are gold and other minerals in the town's slag heaps. They talk of planting pinyon along the contours of the canyon that hugs the town and harvesting the nuts. The old bottles buried in dumps about the town also have a market. In mining towns promise is never entirely abandoned, even after the inevitable bust of the first big strike. These thoughts distract me from my fear and help me finish the drive down the grade, through the narrows, back to the highway beside the lake, and into the town of Keeler. The town has a strange and funky charm, reminding me of the remote enclaves of northern California that were adopted by hippies and artists during the back-to-the-land craze in the sixties and early seventies and that remain stuck in that era by some sort of cosmic glue. Keeler would be an apt place to go if one were trying to flee the tyranny of restrictive suburban codes and covenants.

Keeler's original fifty-two-block plan, which included civic buildings, schools, and churches, has not degenerated to the meager town of fifty souls; rather, its plan was never realized. In 1883, when the town became the southern end of the Carson and Colorado Railway, D. O. Mills, a financier for the line, remarked after riding it into town: "Gentlemen, we have either built this railroad 300 miles too long, or 300 years too soon."[17] Even then, Keeler, like most mining towns, reflected the hurried extractive nature of mining rather than the agrarian ideals of a young expanding nation. A mined landscape is like a room after a toddler has played in it all day: toys dropped and forgotten in hapless heaps, nothing cleaned up, nothing put away—everything left in half-play when the child lost interest or was distracted by some other shiny object elsewhere.

By 1899, the Keeler that writer Mary Austin saw was "a bare huddle of houses beside the leprous-looking crusts of a vague business of commercial salts and borax-making, and an intermittent bottling of the waters from a hypothetical Castilian Spring of supposedly medicinal properties and unimaginable taste."[18] Today the old houses in Keeler decay behind high reeds. The church windows are boarded and broken. Most people there live in mobile homes, one resident lives in a Beakins moving van with wood plank steps built to the freight door. People in Keeler live with the biting flies that swarm the saline lake and hover over the half-full defunct community pool. The depths painted along the blue sides still show above the dark waters left murky with silt and reeds. Front yards are decorated with found glass bottles, antlers, wagon wheels, and little pagodalike glass transformers from the tops of phone poles, creating a montage of Western kitsch and a kind of industrial Japanese garden style. A small trailer surrounded by a white picket fence displays an old totem pole of blue glass bottles. The church, the pool, the town, all are dusty and dismal remains of the Cerro Gordo mining boom.

Even the cemetery, usually an icon of eternal immutability, is tenuous and fragile. Twelve graves remain, buried along a wash, and over the years heavy rains have shifted some toward the lake. Luna Bean (1869–1909) and J. R. Bean (1857–1915) lay side by side with matching white marble headstones, toes to the hills, heads to the lake. Nameless graves with faded headstones, one enclosed by a Victorian wrought-iron fence, share this small cemetery with the Beans. Death reflects culture. The Numu would burn a dead person's house and their belongings "so the survivors could forget their grief."[19] When mourning is done their graves are smoothed over, the dead are let go, and the living continue their migration. Keeler's dead are remembered in this scruffy cemetery between the lake and mines. Like the town buildings, the graves are scattered about in odd disarray.

MARCH 13. Along State Route 190. Crossing over from east to west, looking towards the lake center, where the surface turns to yellow-ish white—the color of an old man's teeth—the desolation, the pause between two worlds, lies flat before me. A road sign warns for cattle, but I would think even a cow would rather be elsewhere. The road undulates with the duned terrain. Large dunes dwindle down to a scraggy beachfront where sand and saltbush are strewn about, dropped out from sheet flows of volcanic stones and pebbles. Bright terra cotta basalt, white tufa pumice, black basalt, some brittle, some hard like glass, a few bits of quartz, and slate-gray lava. One could comb for colors instead of seashells. Further west the barren lakebed fades into farmland and I can hear meadowlarks warble again.

I am a settler, baking bread: Sweet scents from apple blossoms fill the lazy air, muffled hammer sounds sift through on a warm thick breeze, a dog barks. Sitting in the shade of a locust tree, sipping on a tall glass of fresh lemonade, I can hear the gleeful squeals of children playing in the irrigation ditch that runs behind the house. With a contented sigh, I rise, shake the heat out of my long skirts, the sleepy fog out of my head, and turn toward the back porch. The small wood-plank, single-story house seems tiny beneath the towering peaks of the Sierra. Newly planted cypress trees frame the entry. I stop to inspect for young buds, trying to urge the slow-growing trees to hurry up and look more stately. Perhaps I should have planted faster-growing shady willows in the back like my husband suggested, but I always come and go from the back door since it's near the kitchen and thought it deserved to look more formal.

Opening the door sends out a flood of yeasty smells, letting me know I've let the dough rise for too long again. I heft the swollen mass out of the large ceramic bowl I brought from San Francisco on my last visit home and drop it onto the pine table my husband made last winter. In a gasp it deflates to half its size as I begin kneading. Brushing a rogue hair from my face, I let my mind drift away from the monotonous task at hand. As always I think of the home I still miss in Berkeley, the big trees, the gentle sunlight, and the bustling streets on the rare times I crossed the bay by ferry to San Francisco. We moved out here shortly after our last child was born, hoping to build a life in this lonesome valley. Out of our dry sagebrush-speckled 360-acre claim near Independence my husband promised to wrest gardens, orchards, and a house just like the one we left by the bay. In the first years, I prayed he was right as we toiled to rid the land of rocks and brush. Our love grew stronger with each success to push the desert back and make a place for ourselves. Today I feel safe and happy. We have everything we had back home and much more. Most of all we have our home, our land that we have struggled to make work for us, to feed us; we have our freedom.

My youngest son tugs on my apron with an urgent plea for me to let him go with the older children to pick mulberries at the abandoned farm north a ways. Standing just shy of my hip and covered from head to toe with mud, his tiny body tries to swell, like my bread, to impress me with his bigness. Touched by his eagerness to grow, thinking of myself only moments ago, tugging on my cypress buds trying to get them to be more than they are, I smile and nod. Watching him scamper off towards an impatient pack of local children, I yell to my eldest to keep an eye on his brother. I return to my neglected dough, sprinkle it with more flour, hoping to cover another failed attempt. Will I ever learn to bake bread right? Perhaps not everything is within my grasp even when I set my mind to it. Maybe some things never come, no matter how I strive for them against all odds. Pondering this deeper question, I leave the dough and go out to feed the hogs.

This story illustrates the romantic image of the pioneer family toiling with the earth, gaining freedom with every wild acre cleared and tamed. The events that led up to settlement of the Owens Valley by people outside the Numu communities, however, lacked such romance. The pattern of conquest in the valley is similar to those of other regions of the American West: the discovery and extraction of a valuable resource such as gold or silver attracts an agricultural market, which in turn needs military protection from the native population, who are justifiably upset about having their land invaded. The military force subjugates the native people, making further emigration possible. Land settlement is also facilitated by government actions, such as the Homestead Act and Desert Land Act, passed in 1862 and 1877 respectively, to encourage settlement of the far western reaches of the United States.

In 1859, L. R. Ketcham led the first cattle drive through the Owens Valley. He was on his way from Visalia to the Mono mines, where he hoped to sell the beef. By 1861, when the mining districts were going strong, cattlemen brought their stock into the valley to settle. They saw the vast valley dotted with lush meadows of bunch grass as an untapped source of fodder for their stock. What they overlooked, some intentionally and others out of ignorance, was that the valley was already being harvested to its fullest extent by the Numu. Although it appeared virginal to the eyes and minds of cattlemen, who, coming from California's Central Valley, were accustomed to seeing more cultivated and tame landscapes, every acre, shrub, grass, and animal in the valley and mountains that could be used was being used. Cattle destroyed Numu food sources by grazing on *tupusi* and other traditionally harvested plants, trampling soil, and dirtying streams with dung and urine. The Numu protested, saying that no "white man shall . . . sit down in the valley"; by 1862 the conflicts had become violent, and the newcomers petitioned the military to come to their aid.[20]

The military presence and actions in the valley need to be considered within the larger context of California, a new American state experiencing a non-native population growth of over 300 percent in ten years, reaching about 380,000 by 1860. In the Owens Valley, as in other populating regions of the state, emigrants were settling on indigenous people's land, which naturally led to territorial conflicts. The military involvement needs also to be seen within the larger framework of a nation entering into civil war. California, the thirty-first state, was tipping the balance between Union and Confederate states by giving political allegiance to the Union. The populace, however, was not wholly for the Union allegiance, nor were all the governing bodies, particularly in southern California. Only two years earlier, the state legislature had approved a bill that would allow five counties along the southern coastline to become an independent territory. Then, in 1860, only 32 percent of the state's popular vote supported Abraham Lincoln's election to the presidency. On April 12, 1861, the Civil War broke out, the news reaching California twelve days later.

California's abundant natural resources and fine harbors made the state valuable and vulnerable to attack by Confederate forces. Yet most of California's military strength was diverted to the east, leaving the state's land and people unprotected. The California Volunteer brigades were formed to augment the depleted army, and 16,000 men were mustered into service. When newcomers to the Owens Valley requested military support, it was the California Volunteers that came to their aid.

Lieutenant Colonel George Evans of the Second Cavalry, commanding 201 men, made his first expedition to the valley from March 19 to April 28, 1862, traveling 302 miles from Camp Latham, near Los Angeles. On this trip he scouted for a good site for a military post and investigated the extent of Numu hostilities. The emigrants were then escorted out of the valley because Evans was not authorized to leave protection with them. He summarized the conflict in a letter written the day after leaving the valley, stating that the Numu "have 100 or more good guns, and are determined to carry out their threat that no white man should live in the valley," and that the "mining interests of that section are too great for the whites to give it up tamely."[21] He continued in a letter of July 1, 1862, remarking that "without arguing the point as to [the Numu's] right by prior location to the exclusive use of the valley, I will say that it is very evident to my mind that the mines will be of small value unless the valley can be settled and grain and vegetables grown and beef raised to feed miners with."[22] Evans had returned to the valley on June 11 that summer and established Camp Independence on the present-day reservation of the same name. He stayed four months to carry out his orders from U.S. Army Brigadier General Wright to "chastise" the Numu and "protect our people."[23] Evans had taken his instructions to heart, reporting within a month that he was doing as he had been commanded by chastising the Numu "severely" and that since he entered the valley he had "commenced killing and destroying whenever [he] could find an Indian to kill or his food to destroy."[24] Meanwhile, the severe winter in 1861–62 exacerbated hostility on both sides of the conflict.

Evans notes in the July 1, 1862, letter that the army had little success with direct combat because "the valley being open country, without a tree, the Indians can place their lookouts upon the peaks of the mountains along the valley and signalize the appearance of troops for twenty or thirty miles ahead, and upon their approach they can and will scatter into the hills, where it is impossible to follow them." The land afforded protection for the Numu, but their reliance on its spare food resources proved to be their Achilles' heel. Evans adds that the only way the Numu can be "chastised and brought to terms is to establish at least a temporary post, say for one winter, at some point near the center of the valley, from which point send and keep scouts continually ranging through the valley, keeping the Indians out of the valley and in the hills, so that they can have no opportunity of gathering and preserving their necessary winter supplies, and they will be compelled to sue for peace before spring and grass come again."[25] The Numu simply were to be starved into submission. Their ability to overwinter in the mountains was dependent on pinenut crops, which were probably ill-affected by the two-year drought that followed the winter of 1861–62.

Because military resources were directed to fighting the Civil War, the California Volunteers were poorly supplied and Evans' men were themselves nearly starved. For mounts, they were given "the refuse and condemned horses of the First Cavalry" and themselves were "barefooted and naked, for many of them were as destitute of shoes as they were the day they were born, and had no pantaloons, except such as they had themselves made out of barley and flour sacks." Conditions were so severe that Evans said his men were "almost in a state of mutiny."[26] Among Evans' men there were at least five desertions and one suicide between 1862 and 1863. The California Volunteers were Union sympathizers who no doubt thought they would be defending their country from Confederate rebels but instead found themselves nine months without pay, frozen, hungry, poorly mounted, ill-clad, and fighting native peoples to protect, in some instances, rebel sympathizers. Their frustration and discomfort, in turn, was directed towards subduing and killing the Numu.

The order to chastise, which at first seemed to mean punish, evolved in practice into a campaign to completely subjugate the Numu. The protection of the newly emigrated people in the valley became the protection of promise—the promise of mineral wealth and of agricultural development to support the mining industry. The Numu, who had been described as recently as March 17, 1862, as "an inoffensive, gentle race" that was "not very dangerous as an enemy," were soon seen as an obstacle blocking the flow of commerce.[27] The peaceful Numu were then depicted as "wild savages," a "cruel, cowardly, treacherous race," and their land considered "well worth fighting for."[28] A treaty signed in August 1862 kept peace for a while, and the Numu leaders surrendered weapons. From the report, it was clear they had meager firepower: "two rifles, . . . two double-barreled shot-guns, one Sharps rifle . . . and one Colt revolver (large size)."[29] There was no mention of ammunition, but it was likely scarcer than the guns. The treaty was breached soon after, and the fighting and destruction of food continued for nearly another year until, on July 10, 1863, most of the Numu came to Camp Independence and surrendered.

The following day, more than 1,000 native people were taken from their homes and marched to Fort Tejon. So devastating was the exodus, the experience of it is etched into Numu memories and stories. The people were lured to Camp Independence by the promise of "farming equipments, stock, food, clothing and shelter" or "a gift of freshly slaughtered beef."[30] But once at the fort, they were taken to the San Sebastion Reservation near Fort Tejon. Wagons carried some, but "many had to walk" for nearly two weeks to reach the fort. Mary Harry was twenty-three when the people were taken from the valley. Her family lived in the Fish Lake Valley, east of Owens Valley, and had chosen not to go to Camp Independence with the other indigenous peoples, so she escaped the exodus. At ninety-five, living in Big Pine, she recalled what her grandmother had told her of the journey: "These white soldiers had no pity for the poor men, women and children. Whenever they became exhausted and weak and couldn't walk much farther, they, the white soldiers whipped them and very often killed some. If they were killed they would throw the dead corpse to the side of the road . . . Many of the young girls were assaulted and afterwards murdered."[31]

Jennie Cashbaugh, of Bishop, recalls the journey she made as an eight-year-old girl. Her family was among those who came to Camp Independence to get the promised food and clothing. She remembers the way to Fort Tejon:

Some went on bravely, some were too feeble and weak and fell. I saw them lay down to rest or sit down to rest for want of water or food. I saw the white men with long knives stick the knife into their sides . . . Our flesh was to be picked up by the hungry birds and coyotes of the wilds. Fear clutched my heart and mind . . . My poor grandmother sat down for just a second, she was thirsty, she wanted water. Just then one of the men with the swords saw my grandmother sit down to rest. He was upon her in a second and stabbed her through the heart dead. In a pool of blood lay my grandmother, all alone, cold and stiff . . . As small and young as I was I can still see the sight of my dead grandma back there.[32]

Only 850 of the more than 1,000 who began the journey arrived at the San Sebastion Reservation. By January 1864, only 380 people from the Owens Valley remained at the reservation, "almost in a state of starvation"; most had either died or escaped and returned to their homeland.[33]

Camp Independence, having served its purpose, was abandoned. Petition letters (dated September 25 and November 29, 1864) sent to the army by recently emigrated Owens Valley residents pleaded for renewed military protection. A man known as Joaquin Jim had organized a band to aggressively resist settler occupation. At the same time, some settlers were hiring Numu as laborers and refusing to pay them after the work was finished. Camp Independence was re-established under instructions to "protect the settlers in and contiguous to Owen's River Valley, and at the same time to restrain the whites from attacking innocent Indians."[34]

Hostilities continued until the attack of a Numu camp on January 6, 1865. It was a vigilante retaliation for the murder of Mary McGuire, a woman who had run a way station with her husband on Haiwee Meadow, and her young son, Johnny. An entire Numu camp of forty-one men, women, and children was attacked near the shores of the saline Owens Lake. A few tried to swim to safety in water so salty that it burnt their skin before they died. "The last encounter with the Pah Utes took place at Owens lake, where the Indians again murdered a white settler, and on this occasion a score or more of the Indians were forced into the lake to swim for their lives. Most of them were drowned in the soda water of the lake, and it is said that some of the bodies still lie in the bottom, perfectly preserved by soda which has crystalized about them."[35]

After the massacre at the lake, the valley quieted, but the fields that had once provided seeds and roots for the Numu were now run with cattle and sheep or plowed under to make way for crops of wheat, alfalfa, potatoes, and other grains and vegetables. The land was open for settlement by non-native peoples. Camp Independence continued to facilitate settlement and mineral extraction until July 9, 1877, when orders again came to abandon the post.

The Numu, who only a few years before had known every rock, plant, creek, and animal in the valley, found themselves in a strange land. Their territory was lost and, with it, their ability to live by their traditional means and customs. Jennie Cashbaugh, who had escaped with her mother and younger brother along the way to Fort Tejon and returned to the Owens Valley, lived with relatives for a while at Sand Hill near Laws:

> a place where nothing could be raised. We lived there and watched the white man come in and fence the land where we lived and then like all Indians we had to move off the old home grounds that we thought belonged to us . . . What can we do? Nothing but hang our heads in shame and sorrow . . . Nowhere can we gather the seeds and herbs for medicine and food. We have to eat the white man's food entirely, we get sick—the white man's sickness, and get white man's medicine. Our lives are short compared to those of our ancestors.[36]

Staying in one place, building a permanent community village to inhabit year-round, went against the Numu's relationship with the land. A writer of the time describes their condition:

> View your staunch, coppercolored American in the midst of a village, or town, where he stands confused among the mazy ways of civilization, and you see a child, timid, suspicious, curious. There you will find the Indian at his worst, out of place,

out of harmony, a strange discord among humanity . . . The days of pine nuts and beetles have passed for them. The new regime means farm work, day labor, wood sawing, and all sorts of odd jobs which the white man has invented. The Indian does these, not because he loves work, but he has learned the potency of money, and he must work to get money.[37]

Most accounts of the conflicts say the clash was inevitable, that the mightier hand conquered. William Chalfant's perspective in *Story of Inyo* (1933) is typical of the postwar rhetoric, noting that "white domination, and its ability to make use of resources which to the Indian meant far less than the comparative comfort the conquerors have brought to him, were as inevitable here as they have been elsewhere as civilization advanced."[38] Violence and warfare cleared the way for settlers, who then began to appropriate the territory in more passive ways. "In 1866 the town of Independence was beginning to boast permanent buildings, and there were six mills in the valley below Bishop Creek in working order and more under construction, all of which added to the sense of security enjoyed by whites and the realization of the end of their residence in the Owens Valley by the redskins."[39]

After the fighting ceased, many newcomers arrived with minds full of stories of wealth and glory to be had in the valley. Despite the dismal report by A. W. Von Schmidt, who surveyed the valley from 1855 to 1856 and described the region as "worthless to the white man, both in soil and climate," most literature in the late nineteenth and early twentieth century boasted of unimaginable richness and challenge.[40] Even after the mining boom faded, the valley was promoted for its agricultural promise. By 1908, boosters were trying mightily to change the valley's image as part of the Great American Desert to one of a rich piece of frontier California.

The call of the soil is the call to freedom . . . From everywhere comes this call of the soil; from the North, the South, and the West; but the call is loudest and most promising in the new sections which have recently been opened by systematic reclamation, and where great systems of irrigation have changed thousands of acres of arid desert lands into fertile farming regions. Such a section is the Owens River valley, once merely a suburb to Death Valley, but now a paradise of homes, where thousands of thrifty farmers have reaped a bountiful reward from the land of plenty.

The tremendous mining interests, stock raising, lumbering, agriculture, and the vast areas of fertile land which the government is reclaiming through its great irrigation projects, afford a field so attractive, so varied, so rich in prospects, so inexhaustible in its resources, that the home-seeker, the investor or the capitalist will never tire of its exploitation. Practically all the precious metal possessed by this country lies hidden among the mountains of this area, while millions of acres of agricultural lands, hitherto held as worthless, arid desert waste, have been made to blossom by the mere application of water. Here they lie, fields as broad as a kingdom, idly awaiting the coming of the thrifty man who is seeking to make a home somewhere for a growing family.[41]

Families came to the valley hoping to gain the personal freedom that comes with land ownership. Many came from the humid eastern states, the California coast, or the western Sierra mining camps, where rain was more abundant, and they carried with them a common belief of the time, that rain follows the plow. These young settlers often found themselves left high and dry. Early families settled on or near Numu village land along water courses in the lower and middle elevations of the valley, but late-comers had to take dry land higher up on the bajada, where sage-brush was dominant. It was deemed better, even nobler, to home-stead on desert scrubland rather than purchase a tract of tamed bottomland. The process of struggle with the land was believed to make settlers more American. "The first settlers chose homesteads along the creeks and the river. Later settlers had to take land with-out water; but with thousands of swings of their picks and mil-lions of shovelfuls of dirt, they carried the water farther and far-ther from the streams onto the parched land. With the land fertile, and a maze of canals and ditches brimming with water, the 4500 Valley settlers were turning their sagebrush desert into a pastoral paradise of orchards and hayfields."[42]

By transforming the landscape, they transformed themselves; controlling the land gave them strength and independence, and importing familiar landscape forms maintained their ties to the larger America. The metamorphosis of a person through transfor-mation of the land has been described as Americanization: "In first being overwhelmed by nature, and then in overwhelming it, the pioneer underwent a process which 'Americanized' him. It freed him from dependence upon Europe . . . The frontier transformed his old ways into new American ways, and subduing nature be-came the American's manifest destiny."[43] However, this idea of Americanization, ending in the subduction of nature, better de-scribes the pioneers east of the hundredth meridian, where dry farming is possible. On those vast plains, farmers could win some semblance of independence by wrestling with the land. Once the land was broken, farmers could live in relative autonomy. In the arid West, irrigation required some form of social organization to control water distribution; canals needed to be maintained and water needed to be fairly allotted, in wet and dry years. California's pattern of drought and flood has always made this a tricky task.

So, in the Owens Valley, subduing both land and native people was not enough to make the settlers free and independent. The question of how to make the desert reach full bloom still lingered unanswered.

The pioneers settled in, doing what they could to change the desert landscape, to make a living tilling soil or raising cattle or sheep. Draping the strange with the familiar is a common method of making oneself feel at home. Town form and architecture in the valley mirrored what existed east of the territories and in the more settled regions of California. In early times, when lumber was difficult to acquire, people built adobe houses. Along the Owens River near present-day Bishop, Allen Van Fleet made the first structure in 1861, using sod and stone. The 1872 earthquake, however, destroyed adobe structures and made people leery of this building material. Lone Pine was the hardest hit, where "about three-fourths of the buildings were of stone and adobe, and every one of these was dashed into a heap of ruins." No one died in the town of Independence, owing to the timber construction in which "partitions and joists protected the people in the buildings from falling walls."[44] Twenty-six lives were lost in Lone Pine during the earthquake, including a baby killed in its mother's arms, while the only toll was property damage in Independence. Given this contrast, people chose timber to rebuild their homes and businesses. For example, after Camp Independence (1862–1877) lost its original adobe structures in the earthquake, the camp rebuilt keeping the regimental site planning but using timber and familiar plant types. An 1875 circular from the Surgeon General's office notes: "The present buildings are located on the four sides of a parallelogram, forming the parade ground and lawns in front of the officers' quarters. These grounds are set out with trees and covered with grass. A live hedge forms the westerly boundary of the camp . . . Chimneys of adobe [are] built into each house carefully braced

DWELLING BEFORE

with wood and extended above the roofs by galvanized iron flues on account of earthquakes." The camp hospital is "built of redwood lumber outside and in, with valley lumber for frame, roof and floors."[45]

Around the time of the quake, transportation was improving, first with Remi Nadeau's mule train, running from Cartago at Owens Lake to Los Angeles, and then with the arrival of the Carson and Colorado rail line in 1883, which made lumber more available. A typical home used balloon framing, clapboard or board and batten facing, and gabled shake roofs. They were unadorned, small, and simple. Commercial buildings along the main street sported the false fronts that have become so emblematic of the Wild West. Simple picket fences might enclose the homes and civic buildings. Ornamental vegetation and crop plants were brought in from the populated parts of California and from the East. For example, on John and Augusta Kispert's 400-acre spread along George Creek, "when lumber was more readily available a lovely ranch house was built much on the same plans as the Engel home in Minnesota, where Mrs. Kispert was born and raised. Fruit tree seedlings and grape cuttings were sent from Minnesota and planted on the Kispert ranch, giving them a nice vineyard and orchard."[46] Fences up, streets laid out, fields plowed, order established—all marks on the land to say *this is America*.

For the valley, an inescapable consequence of becoming American soil was that the landscape became embedded in economics. Captain Davidson, in his early description, had identified the valley as an ideal place for the farmer, offering "every advantage but a market."[47] The missing market for the farmer had manifested by means of large silver and gold strikes. "The high-priced mining camps of the Sagebrush State have been a very gratifying source of wealth to the farmer of this valley, and every farmer will say so."[48] General requirements for successful Western farming, as described by Frederick Jackson Turner, were well satisfied in the Owens Valley: "Among the important centers of attraction may be mentioned the following: fertile and favorably situated soils, salt springs, mines, and army posts."[49]

Farming and ranching prospered and waned with the mining industry. Crops included apples, peaches, plums, apricots, wheat, corn, hay, oats, barley, and potatoes. Livestock was mostly cattle and sheep, but horses and pigs were raised as well. Other farm goods, such as milk, eggs, butter, honey, chickens, and turkeys, were among the products that went to the mines. Families also grew fruits and vegetables for their own consumption. Ranch life, especially in the pioneering days or during postboom market declines, was a struggle. The daily toil was made difficult by the valley's remoteness. John E. Jones, who came to the Round Valley area near Bishop in 1864, described the early years on his family's farm: "Just think of my wife making candles, sugar, molasses, starch, and even I and the children cutting and burning green cottonwood to have ashes for her to make soap, and I got deer skins to make coats and pants!" Despite the arduous chores of farm living, his image of the valley is one of productive family farms, an Arcadian oasis; "the prettiest valley in California," he called it.[50]

The Shaw ranch in the Bishop area illustrates the diverse population that participated in developing the valley's agricultural land. Clara Shaw Eddy (b. 1903) recalls of her father's ranch the German irrigator, the Englishman who fed the cattle, the Irish handyman, the Swiss man who cared for the dairy cows, the Chinese cook, and the Numu family whose members worked as a cow puncher, a laundress, and a "garden-weeder."[51] On many farms, Numu worked as laborers and became so relied upon that in the 1870s ranchers resisted a plan to move them across the Sierra to reservation land by the Tule River. The move would have depleted ranchers' "indigenous labor force."[52] The Numu also began to

113

homestead their own land under the Indian Homestead Act, passed in 1875.

In the summer of 1900, the Tonopah boom in Nevada drew upwards of 8,000 people to the mines. The Owens Valley supplied most agricultural goods to this population. Cattle production went from 5,997 head in 1880 to 11,119 by 1900. At the same time the sheep count soared from 9,574 to 62,036 head. Milk production rose from 510 to 464,633 gallons, and honey increased from 10,630 to 146,320 pounds. The valley was finally becoming the land of milk and honey that boosters had been promoting. More desert land was cleared, particularly in the Bishop-Round Valley area.

Along with agricultural development, there evolved a local canal system built on the remains of the Numu ditches. The canals off Bishop Creek looked like the spider lines of exposed capillaries on a frost-bitten face. A turn-of-the-century map of this irrigation system shows the waterways running every which way across a neat township and range grid whose squares were designated as cultivated, pasture, or sagebrush land. By 1905 the irrigation system extended for 110 miles and included the community-maintained McNally Canal and the Bishop Creek Ditch, but it had grown piecemeal and lacked organization and efficiency. Enter the Bureau of Reclamation (BOR), a newly formed federal agency whose *joie de vivre* was organization and efficiency. At the request of valley residents, the BOR was considering a project that promised to enlarge the system and expand the irrigable land from the 30,000 acres already under cultivation to 106,000 acres. The BOR surveyed potential reservoir sites and drew 565,480 acres into the public domain around these locations.

A repercussion of the valley's becoming part of the United States and adopting a capitalist economic structure was the eventual dominance of capitalism's nonregional perspective, the kind of long view taken by agencies such as the BOR. In 1893, when

Frederick Jackson Turner presented his paper to the Wisconsin Historical Society on the significance of the closing of the frontier in America, he neglected to add an ample and sustained supply of fresh water to his list of elements a Western farmer needed. At that time, the arid frontier was only sparsely settled and still acted more as a barrier to gold and silver mines in California and Nevada than as viable land for settlement. He did not foresee the immense transformation of the West beyond the hundredth meridian, a transformation brought on by massive federal irrigation and flood control projects governed by the U.S. BOR and Army Corps of Engineers that began in 1902 with the birth of the BOR. The BOR and Army Corps are responsible for changing both the Western landscape and the Westerner's attitudes towards water. The BOR's institutional perspective was that all water left unused by human beings was wasted. The settlers in the valley turned to the agency for help transforming their desert into a more productive farming community. "Forty thousand acres of this irrigable land is under cultivation, while the rest waits for thrifty souls and willing hands. Besides this a recent government reclamation project has under consideration the recovery of 100,000 acres which is now desert, dry, but waiting only for water to spring into a wealth of green fields, rich vineyards and orchards."[53] In retrospect, the settlers might have regretted calling national attention to their remote valley, since the same federal agency from which they sought help ended up assisting Los Angeles in diverting the valley's water and dashing the settler's dreams of rural expansion forever.

115

If this is a desert, what are all these people *doing here?*

REYNER BANHAM, *SCENES IN AMERICA DESERTA*

Lone Pine in the late nineteenth century (*above*). U.S. 395, heading
north to Lone Pine (*below*).

We live in time and through it, we build our huts in its ruins, or used to, and we cannot afford all these abandonings.
WALLACE STEGNER, *ANGLE OF REPOSE*

Mary Lent winnowing in the valley, ca. 1914.

Toni nobe, Numu winter dwelling, (*above*) ca. 1916. Mining shack near Owens Lake (*below*).

Cowboys and over five thousand head of cattle near Independence, 1902.

Independence, below Mount Kearsarge, ca. 1914.

Keeler community pool, near Owens Lake.

Remnants of adobe walls damaged in the 1872 earthquake, Camp
Independence.

Wood frame house in Independence. Timber became the preferred
building material after the earthquake.

Cultivators of the earth are the most valuable citizens. They are the most vigorous, the most independent, the most virtuous, and they are tied to their country, and wedded to its liberty and interests, by the most lasting bonds.

THOMAS JEFFERSON TO JOHN JAY, AUGUST 23, 1785

126

Ira O. Clark standing in a wheat field near Laws, ca. 1915.

Red Mountain Fruit Ranch, part of the thriving agricultural economy around Bishop before increased diversion of Owens River water in the 1920s.

Old Lone Pine cemetery (*above*). Forgotten picket fence in the bottomlands (*below*).

3 LIVES DIVERTED

WAITRESS, MINIMART CLERK, hotel receptionist—the people who do the living and dying in the small towns along Highway 395—when they talk about the Owens Valley, they almost always refer to it *before*. They never bother to explain. "My dad grew up here and he remembers it before, when it was green everywhere, when the valley looked like the San Joaquin," the drug store clerk tells me as she rings up my sundries. With their own point of reference, they mark events in time as either before or after diversion. The Owens River diversion into the Los Angeles aqueduct forever changed how valley dwellers saw and remembered their landscape. Their perspective was forever colored by their sense of loss and injustice. Arcadia was left to die of thirst, and the tragedy of it makes the landscape of recollection more vivid, like a person taken before their time so only their sweetness and promise are remembered. No one was left untouched.

This land before, this remembered place, is more a green garden than a desert. Yet, in 1903, writer Mary Austin had christened the Owens Valley and its surrounding mountains the "Land of Little Rain," the title of her first and most famous book and a name used commonly to this day. She pulled poetry from the brown landscape, where "its treeless spaces uncramp the soul," but she, like many locals, moved on when Los Angeles sapped the life out of the farming business.[1] Perhaps the rich crops coaxed out of seemingly barren soil had made the valley's dry hills more beautiful to Austin. The farmer's struggle against such adversity to spread green vegetation over brown earth and crowd out the desert's natural gray and muted plant life created a lyric of contrasts. Apple groves and alfalfa fields made the surrounding desert feel less threatening, a place to visit, to enjoy, but not one in which to dwell. Austin, who lived in Independence in a small wooden house beneath a willow tree when she wrote her book about the region, always came home to rural comfort after her long wanderings in the desert. Spending days beside the farmhouse or working the green fields, one could forget the sagebrush, the white salty soil by the lake, the burnt volcanic earth, and think only of the apple blossoms' sweet scent and the rippling alfalfa fields of the rural landscape.

Irrigation farmers had made the "desert blossom as the rose," transforming the arid valley into a "land of milk and honey," a feat only possible with extensive irrigation.[2] These bucolic farms could shield settlers from the desert, but they proved no match for the Los Angeles Department of Water and Power (DWP), the U.S. Bureau of Reclamation (BOR), and such men as William Mulholland, Joseph Lippincott, and Fred Eaton. In 1892, Eaton visited the valley at the invitation of Frank Austin, Mary's brother-in-law, to explore a local irrigation project that Austin was considering. Instead, Eaton recognized the potential of the river and its tributaries as a water supply to his hometown of Los Angeles by way of a gravity-fed aqueduct. As it had during the subjugation of the Numu, topography once again was proving instrumental in determining the valley's fate. Keeping the Numu in the mountains had precipitated their starvation-induced suit for peace. Now the 4,000-foot drop from the valley to Los Angeles made bringing water to the city a viable, albeit ambitious, option.

However it wasn't until 1905, when Los Angeles was coupled with drought and growth, that Eaton, at the request of DWP chief engineer William Mulholland, began buying options on property in the Owens Valley. Armed with the BOR maps and survey reports, he focused on land with water rights in the southern half of the valley and the potential reservoir site north of Bishop. He kept his true motives for wanting the land vague, presumably to keep land prices low—some thought he was a cattle rancher, others that he worked for the BOR. The Register at the United States Land Office in Independence, Mary Austin's husband, Stafford Austin, was among those who believed Eaton represented the BOR and that he was securing land for the proposed local irrigation project. Austin had no idea at the time that the BOR had already abandoned the project in favor of Los Angeles's plan to build an aqueduct. When the truth was discovered, he and other valley residents who had no intention of leaving were incensed by Eaton's deception. Those who had sold options to their land became angry because they felt they should have received more money. So began the long conflict between Owens Valley dwellers and Los Angeles. Having already fought the rocky, dry alkaline soil, weary farmers who had not yet sold out had only weak defenses against progress and soon relinquished land and water rights to Los Angeles. In 1913, when the Los Angeles aqueduct was complete, water—the element that bound culture to the desert land—began to drain away, and with it went many valley dwellers' livelihoods.

Valley residents felt duped and resented Los Angeles and the BOR for taking advantage of them. They also resented Mary Austin for not using her command of the pen more diligently to fight off diversion. In 1905, she wrote one article for the *San Francisco Chronicle* criticizing Los Angeles's motives and methods and supporting her husband's claim that Fred Eaton had been deceptive in letting on that he represented the BOR when he purchased the options. Yet, also in 1905, instead of staying to fight diversion, she left the valley, abandoning her husband and leaving her teenaged mentally disabled daughter, Ruth, in an institution, and went to Carmel, then a town with a strong writing community. She left the valley, but not without regret and not without first considering what she could do to stay. In her autobiographical novel, *Earth Horizons* (1932), she reveals her turmoil in choosing to leave Owens Valley. "She walked in the fields and considered what could be done. She called upon the Voice, and the Voice answered her—Nothing. She was told to go away . . . Mary went away . . . She was stricken; she was completely shaken out of her place. She knew that the land of Inyo would be desolated, and the cruelty and deception smote her beyond belief . . . She sold her house in Inyo; she meant not to go there again."[3] Mary left, as did many farmers, giving the valley back to the desert. Austin offered a fictional revision of the story in *The Ford* (1917). In this novel, the heroes—a family who loves the land in their valley, Tierra Longa—not only conquer big city attempts to buy up water rights but also defeat the short-sighted neighbors who sold options on their property. In her novel, as in reality, she blamed both the city's venture capitalists and the valley's disorganized provincials for the loss of the land's water. Perhaps rewriting history consoled her, but by 1917 the valley's water was already flowing from the taps in Los Angeles and Austin had left her Carmel writing studio, built in a tree, to live in New York City.

After 1913, land in the valley's southern half decayed while the northern end continued to prosper. Geographer Ruth E. Baugh, who traveled in the Bishop area in July 1916 described the land she saw as "flowing with milk and honey. From the railroad station at Laws to the town of Bishop and west to Red Hill one traveled tree-lined roads through a district almost continuously cultivated; green fields bordered by wide irrigating canals alternated with stretches of damp pasture lands. Creeks brimful of sparkling water dissected the piedmont sloping to Owens River. A verdant land, Owens Valley presented a scene of substantial economic well-being and human contentment."[4] The demise of the Bishop area began in 1918, when the local water bureau began drilling wells to extract groundwater in response to drought conditions. The alluvium layer, over 7,500 feet deep in places, held vast stores of snowmelt beneath the desert surface. Before pumping, this groundwater was high enough to sustain vegetation even when streams ran dry during drought periods or in the summer months. The DWP began purchasing land in the Bishop area in 1920 and now owns nearly all the land in the valley.

As it had been when the Numu occupied the land, the Bishop area was the most densely populated region in the valley and had the largest amount of land under irrigation, so there the effects of increased water diversion were most pronounced. When Baugh revisited the region in 1936, just twenty years after she had traveled through the thriving rural landscape, she presented an image of devastation: "Land formerly productive has turned to sage brush, and commodious farm buildings are now in ruins. Weed-choked irrigation ditches, abandoned farm machinery, and rows of stark, bleached tree trunks identify the sites of former ranches."[5] Despite the loss of agricultural land use and subsequent depopulation of the valley in the 1920s, the land continued to sustain the people who hung on. During the Depression, because Los Angeles owned most of the valley's land and could afford to pay the property taxes, Inyo County maintained an income during a time when other rural communities were going broke. The strange and often tense relationship between the people of Los Angeles and Owens Valley residents, which began with Eaton's 1892 visit, has left the land simultaneously barren yet full of promise. The bittersweet irony is that the most devastated region in the north has managed to profit the most since diversion.

Some argue that diversion of the river and abandonment of farmland have saved the valley from suburban sprawl, that, if left well-watered, "the valley would become just like Los Angeles—full of people."[6] They claim this remote place would be blanketed with meandering suburbs had Mary Austin stayed and fought with the others, or had valley residents actually won a victory at the Alabama Gates in 1924, when sixty rebellious locals took over the control station at the floodgates for five days and sent the water back to the river. Had they succeeded in keeping Owens River water from flowing to Los Angeles, the desert vegetation and all the wildlife it supports instead would have been sacrificed to bulldozers scraping pads for housing tracts. In defense of its water management program, the DWP now congratulates itself for preserving the beautiful Owens Valley landscape from suburban sprawl. The valley's northern end has been called an accidental national park, a pristine wilderness maintained by diversion. There is some truth to this, but what appears starkly beautiful is, particularly in riparian areas, an ecosystem in decline. If one measures the quality of land by economy rather than ecology, the Bishop area—the region best situated for outdoor recreation opportunities—has recovered the most from the original impacts of diversion. The northern end of the valley now supports a population of 15,000 people.

In contrast, town populations to the south have not changed much since 1913. Few houses are built because the DWP owns most of the land in and around these towns. Structures are moved from time to time—the Commander's house at Camp Independence was relocated to the town of Independence, and some barracks at the Manzanar War Relocation Center were moved and converted into hotel rooms. The aqueduct stopped the valley's boosterish mantra of progress and growth in its tracks. Water diversion magnified the existing desert environment and kept the valley's population from exploding like the neighboring regions of San Bernardino, Tulare, Kern, and Fresno counties. Yet, north or south, despite preservation resulting from diversion, valley dwellers remember a better place *before*.

In 1913, settlers' hopes for a local reclamation project were dashed by the very governmental body they had depended on for help. Their crime? Underutilizing the valley's most valuable resource—water. Conservationist Gifford Pinchot's powerful words, "the greatest use for the greatest good for the most people for the longest time," wrapped into a familiar noose, the same noose the settlers had used to wring the life out of the Numu.[7] It is an ironic twist in perception that enabled the settlers, who had just taken the valley from the Numu, to cry: "Are [our] rights to be trampled under foot? Shall [our] homes be despoiled?—and all without a hearing?"[8] Taken out of context, it is impossible to tell whose rights are being trampled: the settlers' or the Numu's. By the beginning of the twentieth century, settlers believed they had total ownership of the valley and that all its resources were nature's gifts to them. When they first entered, "only Pah Utes roamed the valley of the Owens. The water was here, but it was free, and no one claimed it until the homeseeker came."[9] From the settlers' point of view the Numu had no water rights, despite the miles of canals they built and managed. For water to be claimed, a concept of ownership must be established, a concept that is willingly recognized by the American settlers.

When the settlers came to Owens Valley, they saw a vast unbroken landscape—no fences, no barbed wire, no signs, no claim stakes, no plowed fields. Without these signals of ownership, they presumed it was free land for the taking and ignored the Numu who told them otherwise. Numu villages had specific territorial grounds laid out to include the natural resources each community needed. The territories swept east to west across the valley to include pinyon pine forests, irrigated meadows, and Sierra slopes inhabited by large game animals. None of these lands were fenced. When a pinenut crop in one territory was poor, a neighboring community might share their harvest. Thus land tenure was more an act of guardianship than ownership. The Numu relationship to the land was not exploitive; they took what they needed for nourishment from the environment and not much more. Settlers, seeing this sparse use, did not believe they had stolen the land from the Numu even though there had been no just compensation. Yet they claimed Los Angeles had stolen their land even though property had been purchased from settlers at fair (and not-so-fair) market prices. Perhaps the settlers did not see themselves as thieves because it is impossible to rob from someone who appears to own nothing. They also acted under the support of the federal government, which gave them a sense of propriety. William Mulholland's words—"There it is. Take it"—spoken to a crowd of over 30,000 at the 1913 aqueduct dedication, also could very well have been used to describe the young nation's sentiment towards westward expansion and the subsequent displacement of native peoples.[10] The Owens Valley was theirs, they understood, because their country had claimed the land.

An interesting irony of Western land and water history is Westerners' simultaneous hostility toward and reliance upon the federal government. Like a teenager who rebels against parental authority but expects to borrow the car (and a full tank of gas) on Saturday night, the West wavers between claiming autonomy and demanding aid from Washington. And when government actions are not in the West's best interest, great whining ensues. Without the support of the U.S. military, the settlers would not have been able to take the land from the Numu. They then settled the land with the aid of the Homestead and Desert Land Acts. In 1902, when the Federal Reclamation Act passed (creating the BOR), valley residents once again looked to the government for help, this time to develop the valley's irrigation system. However in 1907, when Gifford Pinchot, who had helped draft the Reclamation Act and was chief forester for the U.S. Forest Service, convinced President Theodore Roosevelt to withdraw land in the Owens River watershed from settlement, valley residents balked. The withdrawl put land that might have been settled under the control of the U.S. Forest Service. This move also protected Los Angeles's future water supply, a fact that was clear to Roosevelt. From his point of view, the valley and its water did not produce enough wealth for enough American citizens to justify leaving it alone. He acknowledged Los Angeles had greater need and potential for greater prosperity than this remote valley hidden behind the giant Sierra Nevada, stating "the opposition of the few settlers in Owens Valley . . . must unfortunately be disregarded in view of the infinitely greater interest to be served by putting water in Los Angeles."[11]

Roosevelt's words mark a pivotal point in the Owens Valley's land-use history; management decisions henceforth ceased to reflect solely the perceptions and interests of valley residents. What outsiders—people who had never visited the valley—believed to be true of the place began to influence the valley landscape. At this point, its location in the Great American Desert became its tragic flaw. To the larger nation, the valley was now fastened to the arid barrier that had for over half a century tormented overland emigrants and kept them from California's riches. The local communities sensed defeat, as a writer for *Inyo* magazine noted in 1908:

The people outside of Inyo who know of this country at all know of it as the land of Death Valley. Death Valley is synonymous with many unpleasant suggestions, and among these is the uncanny view of a great unpeopled waste, latent with mineral riches which the most daring fear to explore . . . Inaccessible, unredeemable, worthless—a mere unsightly blotch upon the fair face of California . . . A land that was given up to horned toads, chuck-wallas and tarantulas, if it possessed a good water supply for Los Angeles, was deemed a proper place to despoil.[12]

Los Angeles's single-minded focus on water left the valley broken, its identity as the land of milk and honey drained away with the river. The green valley seemed to die before settlers' eyes, many of whom had spent their childhood swimming in the canals, watching the river's violent spring floods, and running among shady orchards plucking giant sweet apples from the trees. To them, the land had always been a green rural valley—rich, abundant, and, most importantly, irrigated. In describing the valley before diversion, old-timers inevitably mention water:

Every home had water. You could run all your faucets night and day for $1.50 per month . . . Everyone used this water for irrigating their gardens and trees . . . This water system was a part of the town's economy. Everyone had all the irrigation water needed night and day.

In the early 20th century Manzanar was a lovely rural community. They raised and shipped apples and pears and had large alfalfa fields. The houses were scattered among the trees which hardly needed irrigation as the water level was high from the run-off from the hills . . . An apple orchard full of

LIVES DIVERTED

pink and white blooms, with a background of snow-capped Sierras is a picture I shall never forget. But the City of Los Angeles needed more water! . . . The people moved away. The houses were moved or torn down. The orchards were made into firewood and alfalfa fields slowly dried up and desert shrubs reclaimed the land.[13]

Without water, Mary Austin's brown land became browner. Formerly irrigated orchards and fields withered as desert vegetation spread along the valley floor. After more than forty years of being cleared of desert species to make way for agriculture, the valley bears scars that have yet to heal. Since 1913, rabbitbrush and saltbush have rooted in fallow fields. These opportunistic species inhibit diverse plant communities from re-establishing because they thrive and propagate more readily than other natives. Aggressive exotic plants such as Russian thistle, tumbling symbol of the lonesome West, also crowd out indigenous species. Yet the desert returns with slow and patient deliberateness. While the return of native vegetation could be seen as a rebirth of the desert, families remembering the once green valley regard this initial reclamation as one might witness the death of a loved one.

Our family knew Owens Valley in its primitive state. We saw it gradually fade from a paradise of wild game, stock raising, orchards, and fields of alfalfa to a hopeless desert . . . No one now would suspect that those miles of brush wasteland were once beautiful Wild West farms nestled along the foot of the snow clad Sierras.

One can deal with bereavement, anger, the terror of cataclysmic events, but to sit in the midst of a cherished dream and watch its gradual decay was like the anguish of a crucifixion that continued day after day without end and without hope. With each plant that withered, so did some of the zeal that they had brought to this mission field.[14]

Genny Schumacher Smith, editor of *Deepest Valley* (1962), a compilation of natural and cultural history of the Owens Valley, summarizes the sentiment of residents as more water flowed into the aqueduct: "The '30s were sad years for Owens Valley . . . Its ranches turning brown and only small mines operating, the eastern Sierra's future seemed bleak indeed. Forlorn houses and barns with dead trees, weed-grown fields, neglected fences, and empty ditches were poignant reminders of shattered hopes and plans. As one pioneer, returning for a visit, put it, 'Every time we pass the old home place, my wife cries and I swear.'"[15] People born and raised in the valley, who had only known joy and plenty from the land, were heartbroken as they watched all they had worked towards turn to dust. Many moved away. Those who stayed on had to adapt to life without irrigation. Some caught the early scent of a tourist market in their beautiful landscape and established mule-packing companies, restaurants, motels, or coffee shops. Others applied their cowboy skills to the silver screen by working for companies that came up from Los Angeles to film the increasingly popular Westerns.

> In the hopeless years when their ranches were drying up, some Owens Valley people began to realize they still had the richest bonanza of all—their magnificent mountain and desert scenery. Today tourist business is double the value of ranching, mining, and lumbering put together.

The Dow Hotel in Lone Pine was built in the early 1920's. At that time Hollywood was looking for locations where there was a variety of scenery. Where better than the Owens Valley, with its snow-capped Sierra, its ancient Alabamas, its deserts and mountain lakes and streams . . . By building the Dow Hotel there is no doubt that Mr. and Mrs. Dow are responsible for much of the prosperity that came to the Owens Valley when the City of Los Angeles bought out the land and water.[16]

Trout and mules helped to ease the shift from agricultural to recreational land use in the valley and surrounding mountains. Trout fishing along a high mountain stream is the old-fashioned way of communing with nature, of having the ultimate outdoor experience. Although the waters in the lower valley were heading to Los Angeles, the high mountain streams still ran free and were thick with trout. Standing on a flood-polished boulder, watching the fishing line drift down rapids past a quiet pool, just waiting for a hungry trout to bite, became a leisure pastime. Trout fishing was not a new sport for Owens Valley residents. By 1875, the Kern Canyon area near Golden Trout Creek was known as *the fishing grounds*, and a man named Dick Runkle had established a pasture and store there. He sold fish in Lone Pine each week, and valley residents often came into the Kern Canyon to fish.

Along the Kern River I was taught by my dad and uncle to catch trout with salmon eggs, a clumsy but hypnotic method, an easy excuse for sitting by a river and watching the water flow by without having to work too hard at much of anything. Trout tend to huddle in calm waters behind large boulders, waiting for bits of detritus to wash by in the rapids. An unsuspecting or inexperienced fish will dash out after a salmon egg as it drifts by and wind up dangling at the end of the line. As a kid, I found that trout fishing was my way of being on my own. As long as I carried a pole and a jar of salmon eggs, I was free to wander up and down the Kern alone. Back then, catching legal-length trout with an old

142

bamboo pole was easy even for a skinny ten-year-old girl. I would imagine I was out in the middle of the wilderness, on my own, hunting for my supper. This was nearly true, but I could always return to camp empty-handed and enjoy my dad's delectable dutch-oven-baked bread. Angling ethics changed as the waters became fished out, and the barbed and baited hook I once used is now frowned on by fishing enthusiasts. Nowadays it's better for an angler such as me to just leave the pole at home and watch the river unencumbered, since the catching harms the fish and not all are keepers.

Back then, I knew nothing of the Mount Whitney Fish Hatchery, of air-dropped fingerlings, of the men who packed into the Sierra each year to collect golden trout eggs. Just north of Independence at the fish hatchery, an odd Swiss chalet structure, trout are still raised to replenish Sierra mountain and Owens Valley streams and lakes. Twenty-two million eggs are harvested each year from the 11,000 brood trout at the hatchery. In the early days, fingerlings were distributed by truck, mule, and rail. Nowadays, every July, two million fingerlings are dropped from planes into backcountry waters. The hatchery land was purchased by the people of Independence, and construction began in 1916, just three years after the aqueduct diverted water to Los Angeles. It was thought fishing could help offset some of the economic loss caused by diversion for this former farming community. And, just as citizens hoped, trout fishing in the Sierra has now become one of the draws to the Owens Valley. Angeleno anglers flock north during the season to fish the waters that cascade down the mountain canyons or along the remnant flow of Owens River. Today, there are more tributaries to the Los Angeles aqueduct than to the Owens River, so the river itself offers only mild adventure for any hip-wader-clad fisherman casting flies across its waters. Rainbow, brook, and brown trout can be found in most streams and lakes in and around the Owens Valley, and all are raised in hatcheries. Jim Harvey, the head packer for the Sequoia National Park, called this trout trio the *triple crown* when he told me that he once packed a photographer into the Sierra who swam the streams and rivers in scuba gear just to capture the three species on film.

Golden trout have proved a more elusive species, since they cannot be spawned in captivity. In the Sierra, the story of the golden trout, California's state fish, began 20,000 years ago. Goldens, the prettiest trout in the region and the only fish not found elsewhere, are thought to have evolved from rainbow trout that moved in from the Pacific Ocean. During the ice age, the mountain streams became locked in by glaciers and all rainbow populations perished except those in the Kern and the Little Kern waters. These early survivors evolved into the Little Kern golden trout. Isolated populations later avoided hybridization with rainbows that migrated from the Pacific after the ice melted and high Sierra streams were once again opened to the sea. High glacial lakes barren of any fish were planted with goldens some time after people began to occupy the area.

It's a matter of debate as to who was the first to plant these high waters with trout. Some believe indigenous peoples who lived in the mountains, perhaps the Miwok or Tübatulabal, were the first to seed waters in order to increase their fish supply. Others believe fish seeding began when nineteenth-century explorers, military personnel, and emigrants entered the region. Cottonwood Creek, an east-flowing stream that spills into the Cottonwood Lake system within the eastern range, was planted in 1876 by Colonel Sherman Stevens. This is the most accessible population of goldens in the Sierra. At first snowmelt each year, an egg-collecting camp is set up by the Cottonwood Lakes. Eggs are packed out on mules and taken to the Mount Whitney Fish Hatchery. The golden fingerlings that are raised at the hatchery are later released from the air. This artificial life cycle, including mule trips and plane rides, maintains the fishing industry and supports the illusion of a wilderness experience for urbanites from Los Angeles and other visitors from around the world. Human intervention has seeped deep into the land, into its waters, into its fish.

Planes and mules also brought tourists into the wilderness when Owens Valley residents began to sell their scenic landscape instead of marketing alfalfa, corn, potatoes, apples, peaches, plums, and cherries. For a time, on the Kern Plateau at Tunnel Meadow, a landing strip and cabins accommodated tourists flown in from Lone Pine. Despite the romance of landing a small prop plane in a remote highland meadow, mules have a longer legacy linking valley dwellers and visitors to the western mountains. The mule, frumpy but sure-footed cousin of the horse, helped forge a lasting bond between mountain and valley. Upon their strong backs, the weight of trailblazing, improvement, and maintenance continues to rest. In 1862, John B. Hockett began building a trans-Sierra trail to link Visalia to the Coso mining region. Mules no doubt worked this trail, which became a toll road when it was completed in 1864. Toll rates were fifty cents for a mule or horse, twenty-five cents a head for cattle, five cents a head for sheep or goats, and twenty-five cents for a person on foot. The trip from the Coso mining region to Visalia took three or four days by horse or mule. When a route up to Mount Whitney was found, mules were used to haul concrete and railing to make the climb safer for tourists. This improvement was funded by the citizens of Lone Pine and completed in 1904. Mules also packed in, piece by piece, the French-designed bridge that spans the Little Kern River. The steel I-beams were laid across the backs of two mules who walked side by side from the end of the road at Mineral King, a western point of entry, to the southern Sierra.

A mule is the offspring of a jackass and a mare and comes away with the best (and sometimes worst) traits of both. A similar animal, called a hinny, comes from mating a female donkey, a jenny, to a stallion. This is less common, perhaps because of the risk to the small-framed mothers during gestation, labor, and delivery. Mules have the nimble feet, strong back, and loyalty of the ass and the height of the horse. In *Mineral King Country* (1988), Henry Brown says mule packing "is a Spanish art form that was perfected during the mining and ranching of colonial Mexico."[17] He notes the Mexican influence in the Owens Valley area, giving the example of the Olivas family, who grazed cattle at Monache Meadows on the Kern Plateau and have run packing outfits since coming to the valley in the late 1880s.[18] Frank Olivas came from Mexico, leaving Sonora in 1885 on horseback with his sons following on burros, "finally settling in Lone Pine where he was a packer and a miner." His son Carmen was also a packer, and his grandson Henry was "a packer, a cowboy and cattle rancher all [his] life."[19]

Along with the oxen, mules bore the burden of the emigrant flood across the plains, deserts, and mountains into California in the mid-nineteenth century. Without mules, many a Western pioneer would have been left scratching his head in Missouri, wondering how the family and all their belongings would make the overland trip to the Western territories. The first wagons to travel in the Owens Valley were pulled by mules. Pioneers toiling to cross the Sierra Nevada found mules were often the only way to negotiate the steep and rocky terrain. Sometimes the animals even became dinner when provisions were low and game was scarce.

Those who have spent time with mules gain a fondness for their faithfulness and an appreciation for their strong backs. Their loyalty, doe eyes, and nearly inexhaustible strength fasten them to their owner's hearts like faithful dogs. Clarence King, in his recollections of traveling in the Sierra Nevada, found them to have mys-

146

teriously seductive and endearing powers akin to feminine allure: "There are certain women, I am informed, who place men under their spell without leaving them the melancholy satisfaction of understanding how the thing was done. They may have absolutely repulsive features, and a pretty permanent absence of mind; without that charm of cheerful grace before which we are said to succumb. Yet they manage to assume command of certain. It is thus with mules."[20] He speaks of women here in a secondhand way, so it seems that in his lifetime he was more beguiled by mules than by women. Such is the charm of mules. John Crowley, who in 1947 started packing in the southern Sierra at age fourteen, summed up the mule in the concise way that men who have spent their lives in the mountains tend to speak; "There are good mules and bad mules, but no dumb mules that I have ever known."[21]

Traveling into the mountains from Owens Valley with a pack train is a different wilderness experience than backpacking. In my great-grandfather's time, during the early days of the Sierra Club, trips into the mountains were almost solely done with pack mules. Mules can carry considerably more than a person, so packing light was not such a concern as it is today for the backpacker. Except for John Muir, who often embarked by foot on mountain trips carrying a handful of tea bags and a wool blanket, few travelers ventured into the Sierra Nevada without many homey comforts. And if something was forgotten, it might well be purchased along the Kern River at the store in Lewis Camp, which sold goods to tourists, packers, and mountaineers from 1875 to about 1950.

Backpackers, however, needn't worry about where to find the next grass along the trail. When traveling with stock, one moves from meadow to meadow so the animals have good pasturage. Nowadays, because of increased sensitivity to the ecology of these meadows, stock use is carefully monitored. Packers like Charley Morgan, whose family used to own the Mount Whitney packing outfit in Lone Pine, take precautions to protect the meadows from overgrazing. He said in a 1993 interview:

> I would like to think that all the early packers had some knowledge and interest in conservation. Where we had stock, we would camp in areas that could handle them. It was responsible use, is what it was. We would never camp that many people in an area where there was a lot of pressure. In other words, we wouldn't camp a Sierra Club High Trip at timber line. We would camp where it was ecologic, off the beaten path . . . The Sierra Club was conscious, and we were too, that we didn't leave any impact. We were continually getting naturalists to go with us to monitor what was going on.[22]

On a recent trip, I passed twenty-three named and countless unnamed meadows in ten days of packing on the Kern Plateau. The trip was the reverse of a route my father had taken with my great-grandfather long before I was born. My dad had been forced off the Plateau and into the Kern Canyon by the mosquitoes that swarm the wet meadows after the snowmelt soaks the soddy soil in spring and early summer. I traveled in the dry late summer months, when afternoon thundershowers evaporate quickly on sun-baked granitic soil and the meadows lose their sogginess. Yet the soil still smelled sweet and fecund in the meadows, which are like the very wombs of nature on the high plateau. These meadowlands—Summit, Long Stringer, Templeton, Movie Stringer, Ramshaw, Volcano, Little Whitney, Junction, Rock Creek, Crabtree, and others—are the same pastures that fed sheep on their route from the Central Valley up through Owens Valley to mining country and also served as summer cattle graze.

The sheep are gone, but mules, trout, and cattle linger on, weaving the cultural landscape of the valley with the high wilderness of the Sierra Nevada to create a tapestry blended with rural fields, meadows, and wild mountain streams. These animals all help maintain the valley's link to the land before. Cattle seem to be the only beasts who continue to stir up controversy. When water was drawn off Owens Valley land and into the aqueduct, cattle replaced crops and orchards. And long before diversion, their wanderings were the first cause for human conflict in this area during the nineteenth century. Their presence in the West has never been resolved. It remains as unsettled as the dust kicked into the air during a stampede. Cattle are emblematic of the Wild West, but they helped drive the Numu to starvation; their grazing has transformed the land, spread invasive plants, and caused gullied erosion. Their roamings on the dry range of the valley bottom and summer wanderings on the Kern Plateau have never been without conflict. And in the end, they become hamburger meat.

Filmmaking is another industry to emerge from the wreckage of diversion. Through film and later television, the perception of the Owens Valley landscape as the Wild West endured in image, if not reality. According to Dave Holland, author of *On Location in Lone Pine* (1990), filming may have started here as early as 1919, with a Will Rogers movie ironically entitled *Water, Water Everywhere*. More than 150 films, mostly Westerns, feature scenes shot in the valley. The Alabama Hills were a popular location because of their interesting rock formations, which made excellent hiding spots for an ambush. The Owens Valley has played the role of Arizona, Nevada, Alaska, Oregon, Utah, Colorado, Texas, New Mexico, Wyoming, Montana, and even such far-off places as India and the South American Andes. Although the genuine lonesome cowboy was a dying breed in the valley, Hollywood's presence kept the mythological character alive for moviegoers and local youngsters. "I lived, went to school, and grew up in Lone Pine. I had many thoughts of being a great cowboy, inspired by the movie outfits that were taking pictures for their movies in the Lone Pine area, Alabama Hills, and great Eastern Sierra."[23]

The cowboy, who finds freedom in his wide and wild West, where the air smells of sagebrush and gunsmoke, lingers on more in legend than in reality, yet this romantic image continues to influence land-use decisions west of the Mississippi River. In 1991, the *Newsweek Magazine* September cover story, "The War for the West," featured information on resource conflicts in the West and examined the cowboy controversy. Writer Bill Turque addressed why ranchers continue to enjoy government privileges such as tax subsidies and access to federally owned land. He suggested "one reason is the aura of the cowboy myth, which still plays in Washington" and went on to quote David Alberswerth, public lands director of the National Wildlife Federation: "They look great, coming [into committee hearings] wearing boots, silver buckles and hats. They're very entertaining and very forceful."[24]

Hollywood has helped perpetuate the mythic cowboy, imbedding him and his landscape deep into countless hearts. I imagine many Americans, when asked to name a cowboy, conjure up the Lone Ranger, the Cartwrights of television's *Bonanza,* John Wayne, and Hopalong Cassidy. All were fictional characters or actors for film and television who spent time on sets in the Owens Valley. The valley became a cowboy's landscape in the memories of people who never saw the real place, never smelt the sagebrush after a summer thundershower. John Wayne, in particular, personifies the Great American Cowboy, and in Lone Pine he is practically a god, his picture hanging on every wall in just about every restaurant, motel, and gift shop. He made his starring debut in the valley, playing a scout for a wagon train bound for Oregon in *The Big Trail,* a 1930 epic about the pioneer tribulations along the Oregon Trail; it is fitting that he also made his last film appearance (a Great Western Savings commercial) near Lone Pine.

Two Westerns, one starring John Wayne and one starring Steve McQueen, feature the valley as a desolate, waterless wasteland where men wind up wandering lost without horses, guns, or drink—the three essential ingredients that give grit to the cinematic cowboy. In *The Three Godfathers* (1948), Wayne's character, Robert Marmaduke Hightower, drolls to his outlaw partners "it looks like you and me are going to chew a lot of barrelhead cactus" when they reach a tank only to find some tenderfoot had dynamited it in a foolish attempt to draw water. Hightower proceeds to squeeze water from the cactus to save one of his bank-robbing partners who had been shot, and he later gives the liquid to a newborn whose mother dies from postpartum blood loss in a covered wagon beside the dry tank. With her dying words, she leaves her baby in their care. In reality, barrel cactus juice causes vomiting and dehydration, but it is the cliché act of desperation to drink it when faced with dying of thirst in a Western desert. McQueen's character, Max Sand, faced with dying alone in the desert in *Nevada Smith* (1966), tears apart an agave to drink from the water that has pooled in its heart. By 1966, Hollywood was embracing technical accuracy a bit more than in previous decades; drinking water trapped in agave leaves, a tactic dependent on the presence of the plant and a recent rainfall, is at least safer than sucking cacti if one is trying to survive in the desert. In both films, the desert becomes the foil against which rugged men demonstrate their raw resourcefulness and fortitude.

Beyond providing big ambush rocks and sandy deserts for Westerns, the valley's desert—where the landscape is so empty just about anything is imaginable—has starred in science fiction films such as *Star Trek V* and *Tremors,* both released in 1989. In the former, Captain Kirk goes on a would-be rescue mission to Nimbus III, a "worthless lump of rock" where the "dregs of the galaxy" mingle in a dark and dusty bar in Paradise City, set along the dry shore of Owens Lake. In *Tremors,* two average losers become heroes in the town of Perfection when they defeat giant people-eating sand worms that have turned the valley into "one long smorgasbord." The thick alluvium that blankets the valley floor makes an ideal habitat for these sci-fi creatures that suck people—and their station wagons—into the sandy soil.

Little evidence remains of the film sets, except for a few old roads built by film companies. In the Alabamas, near the *Gunga Din* (1939) location, an old trash heap spilling down a ravine harbors rusted beer cans and empty ham tins. Food cans of this vintage suggest Cary Grant and Douglas Fairbanks, Jr., may have made supper of the contents after a long day spent playing elephant cavalrymen. Locals take pride in the valley's attention from Hollywood, and each October its fame is celebrated at the film festival in Lone Pine. But the Owens Valley's acting career has done little to change the valley's face. Hollywood is not the greatest landscape transformer to come to Owens Valley from Los Angeles; this role is still played by the municipal Department of Water and Power.

As a teenager, my first impression of the Owens Valley was of a rural environment that had been wronged by a big city. My high school geography lessons were spent peering over a plastic relief map of California while my teacher ran the fingers of his left hand along the long Owens Valley. He was missing a thumb, so his hand always held a fascination for his students, and the movements of his fingers are still memorable to me over twenty years later. As a sophomore who'd never been to the valley, only bits and pieces stuck with me . . . Los Angeles was bad . . . this once beautiful rural valley has been destroyed by greed and politics. This story of Owens Valley water extraction has inspired environmentalists and celebrities to defend the region from the DWP and has spawned ecotours to allow nature lovers to photograph wildflowers in the

Onion Valley or watch tule elk graze by Tinemaha Reservoir, one of the artificial water bodies formed to hold water that feeds the aqueduct. Ecotours bring urbanites to environmentally endangered places either to work directly in conservation and protection or to become enamored enough by the place to take the cause home with them; in this way, the valley is finding a new market.

Tourism, to some extent, particularly here, connects a person to place; but because of the fleeting nature of the modern tourist, the landscape must be reduced into sound bites that can be digested during a weekend getaway or a side-venture during a ski vacation. Tourist experiences are now designed to be superficial so they can be fully comprehended quickly by a stranger. A facade of a healthy landscape, like false fronts in a mock Western town, might be enough to satiate the weekend tourist. Can these experiences inspire strong and lasting love, the kind of love that nurtures enduring protection? Or will most visitors lose interest after returning home and becoming distracted by work, family, or some other environmental cause?

Between Lone Pine and Independence lies a patch of land that is particularly steeped in history and has enjoyed just about every type of human use in the valley. Though the land holds no stores of ore, one could mine from its stratum all the land's tales; every group of inhabitants seems to have lingered for a time within its borders. The Numu, who called it *Tupusi witu*, meaning *ground nut place*, used to camp beside its streams while collecting the bulb plant that is also known in the valley as *taboose*. Nineteenth- and twentieth-century settlers found it to be a rich area for apples, giving it its current name, Manzanar, from the Spanish word *manzana*, meaning *apple*. The old town plat shows the hopeful plans of farmers, with its system of concrete irrigation pipes laid out to help turn the desert into an agrarian paradise. Martha Mills, granddaughter of a Manzanar pioneer in the late nineteenth century, recalls: "Manzanar was a very happy and pleasant place to live during those years, with its peach, pear, and apple orchards, alfalfa fields, tree-lined country lanes, meadows and corn fields. There was adequate water from Shepherd and Bair Creeks, and George's Creek supplied water for the larger ranches on the south . . . Those with larger orchards sold their fruit to markets in Los Angeles."[25] Los Angeles bought the property for Manzanar's water rights soon after the turn of the twentieth century, letting the orchards die and the land return to desert. Until 1942, it was a typical tract of valley land with a typical land-use history—typical, that is, until the United States government designated it as one of the eleven Japanese American relocation camps that interned 114,490 people from 1942 to 1946.

The original plan was to send all people of Japanese descent in the Pacific Coast War Zone to the Owens Valley, but the DWP squawked about all the water it would take to provide for this population. In the end, 5,700 acres were designated for the Manzanar War Relocation Center, a place that became home for more than 10,000 evacuees. Walking among the camp ruins, one may wonder how such a place ever came to be on American soil just a short time ago. At the time of the Pearl Harbor bombing, nearly all Japanese immigrants and their Nisei children (American-born children of Japanese immigrant parents) lived along the Pacific Coast. Although Japanese immigrants were not allowed to naturalize and were thus barred from owning land, they had begun to purchase property under the names of their American-born children. Bringing a long agricultural legacy from Japan, by the beginning of World War II they were in strong competition with other California farmers.

Some argue that the total relocation of Japanese Americans had more to do with eliminating agricultural competition than with ridding the coastal region of potential disloyalists. The lack of actual criminal behavior by the relocated Japanese Americans supports this view. As Dr. Eugene V. Rostow puts it, "One hundred thousand persons were sent to concentration camps on a record which wouldn't support a conviction for stealing a dog." The other perspective, held by people horrified by the attack on Pearl Harbor, saw all Japanese as potential enemies because "there was no possible way of separating the loyal from the disloyal."[26] Racial animosity became ubiquitous in California. A woman living in Lone Pine explained it to me simply: "We were taught to hate [the Japanese] as children in schools, by our teachers." She also held the common belief of the time that the camps protected the Japanese from racial violence. "The kids," she said, "they would be beaten bloody in the cities. They were safer in the camps." Ritsuko Eder, who was interned at Manzanar, held this view as well: "I personally felt that since it was so hard for most people to actually identify Japanese from Chinese or whatever, that it was up to us to cooperate, because in many ways we would be protecting ourselves also to be isolated instead of just being free."[27] She gave birth to her son at Manzanar.

Whether the decision to remove Japanese persons from the Pacific Coast was predicated by greed, fear, or hatred, it made their relocation no less tragic, as many innocent immigrants and American citizens—old men and women, children, and babies—were sent to live behind barbed wire, stripped of their right to freedom. It's predictable that many war relocation camps were sited in deserts. The American deserts are a storehouse for human fears, places in which to prepare for war. The largest military bases in the West are in deserts. The remote and harsh land that surrounds these bases provides a barren, eerie setting that fuels the endless rumors of alien invasions and government conspiracies that haunt places like Area 51 in Nevada or the China Lake military unit just south of the Owens Valley. Deserts are feared for their strangeness, their blatant indifference to human life—apt places to hold people who are misunderstood and feared. Like some earthly purgatory for sinners, deserts make good homes for society's secrets and society's shunned. Over half the people interned during the war were held in desert or arid centers—at Poston and Gila River in Arizona, Topaz in Utah, Granada in Colorado, Minidoka in Idaho, and Manzanar. In all, the centers "were much alike in their isolation, rugged terrain, primitive character, and almost total lack of conviences at the start."[28]

Numbers alone begin to convey the enormity and intensity of the operation. At Manzanar, 105 tons of nails were used to build the flimsy barracks that were soon filled with families. A typical barrack housed four families in its 20-by-100-foot structure. Thin partitions divided the space into four to six apartments. The barracks were lined up two by two to form blocks. A typical block had fourteen barracks, a mess hall, a recreation hall, separate men's and women's lavatories, and a laundry and ironing room. The blocks were set on a grid and numbered one through thirty-five. At the peak of occupation, Manzanar had 10,046 internees, an average of twenty people living sardinelike in every 2,000-square-foot barrack, with 280 people per block lining up to use twenty-four toilets. Typical of military latrines, there were no partitions, just a double-loaded line of free-standing toilets; modest old women toted around cardboard boxes for privacy. Four-hundred-seventy-nine babies took their first breaths at Manzanar.

Evacuation happened quickly, so Manzanar was hastily constructed and accommodations were incomplete when people started to arrive; some barracks were up, but they had no doors or windows and only army cots for furniture. Internees salvaged scrap lumber from the camp construction and built furniture themselves.[29] When evacuation began, there were not enough barracks or materials to build partitions for one-family apartments.

This crowding and lack of privacy further disoriented these up-rooted people. The land and climatic conditions were completely foreign to them, as most had lived only in coastal areas. Most recollections of the early days mention dust. It worked its way into barracks, into food, bedding, clothing, in people's eyes, ears, nose, and between their teeth. The dust was so severe that the Soil Conservation Service developed a plan for the Center that recommended planting 21,000 trees and 25,000 shrubs to ameliorate the dust problem. George Fukasawa recalls arriving at Manzanar: "We got there right in the middle of one of those windstorms that were very common in Manzanar. The dust was blowing so hard you couldn't see more than fifteen feet ahead . . . Everybody that was out there had goggles on to protect their eyes from the dust, so they looked like a bunch of monsters from another world or something. It was a very eerie feeling to get into a place under conditions like that."[30] Yoriyuki Kikuchi, a dentist who was interned, remembers that "when the wind blew, it was terrible, just like Imperial Valley sandstorms. Oh, everybody resented being put in such a place, especially when they were suffocated by sand!"[31]

Not all internees' first impressions were so negative. Water-colorist Kango Takamura found the landscape "beautiful. All pink. The mountains around there were all pink. So beautiful. Yes, I thought this is such a nice place."[32] Artistic responses to the land through garden design, watercolor painting such as Takamura's, and writing seemed to help the internees process their experiences at Manzanar, to reconcile themselves with the mountains and desert that were so strange to them. In her poem, *Manzanar,* Michiko Mizumoto, a fifteen-year-old when she was at Manzanar, reveals the mountains as a source of strength, of maternal security, yet also sees them as part of the strange environment in which she was confined:

Dust storms.
Sweat days.
Yellow people,
Exiles.
I am the mountain that kisses the sky in the dawning.
I watched the day when these, your people, came into
   your heart.
        Tired.
        Bewildered.
        Embittered.

I saw you accept them with compassion, impassive
     but visible.
Life of a thousand teemed within your bosom.
Silently you received and bore them.
        Daily you fed them from your breast,
        Nightly you soothed them to forgetful slumber,
Guardian and keeper of the unwanted.

They say your people are wanton
        Sabateurs.
        Haters of white men.
        Spies.
Yet I have seen them go forth to die for their only country,
Help with the defense of their homeland,
America.

I have seen them look with beautiful eyes at nature.
And know the pathos of their tearful laughter,
Choked with enveloping mists of the dust storms,
Pant with the heat of sweat-days; still laughing.
        Exiles.

And I say to these you harbor and those on the exterior,
"Scoff if you must, but dawn is approaching,
When these, who have learned and suffered in
	silent courage;
Better, wiser, for the unforgettable interlude of detention,
Shall trod on free sod again,
Side by side peacefully with those who sneered at the
		Dust storms.
		Sweat days.
		Yellow people,
		Exiles.[33]

Little remains of the camp structure. The camp's barrack architecture, laid out on a grid of streets, was decidedly military, but only one building, the high school auditorium, still stands. The barracks were sold to World War II veterans and their young brides after the camp had been dismantled. A few were bought by locals and moved to town; some now serve as motel rooms. The roads that divided blocks are fading into the desert soil. Now these empty roads connect nothing to nothing, straight versions of the lost winding mining roads that spider up the Inyo. More striking than this grid of crumbling asphalt roads is the careful placement of granite stones that encircle trees, line walks, and create a garden feeling even in the sparse ruins. It's fascinating to see the remnants of Japanese internees' marks on the landscape; the granite feels charged with history—touched by human hands. On each visit I resist the temptation to carry a rock home with me. The stones were brought down from the Sierra to ring the trees and line the walks for the bustling, crowded community. Imagining the labor makes tangible the already pungent isolation of the camp ruins.

MARCH 26. At Manzanar War Relocation Center ruins. It's ironic that Manzanar shares the valley with a town called Independence, a town so-named in honor of the Fourth of July, a day celebrating freedom for all Americans. The ringing silence and solitude found here today speaks nothing of the confined bustle of 10,000 Japanese Americans held here during World War II. In the camp cemetery a whitewashed obelisk bears three Japanese characters meaning *Memorial to the Dead* and stands as tall and still as the majestic Sierra Nevada peaks. On the monument people have begun to leave found objects as offerings—a mayonnaise jar holding water and three granite stones, the label faded and peeling away, a piece of granite sparkling with mica in a small aqua wicker basket, a chunk of concrete and a rusted scrap of metal, many pennies, broken dishes, granite stones, old ceramic plumbing shards, glass, obsidian, bits of a child's plate decorated with a bear, giraffe, and monkey, a child's plastic sword, a small granite stone wrapped in Japanese newspaper, a wire bent into a heart, bits of abalone shell, remnants of brick and concrete, a piece of rock the color of gunmetal, a bullet wrapped with wire, three small keys on a leather strap, an unopened can of Coors, dried flowers, three dimes, a 100 yen coin, broken clam shells, green glass, sandstone and quartz, a dried thistle stalk, folded origami bound by wire garden-ties to a broken metal handle, a rusted tin can holding a stick wrapped in wire, and rusted springs. Tokens perhaps left by camp survivors and their descendants who make a pilgrimage to Manzanar each year in April.

The whitewash peels from the obelisk. Concrete bollards wrapping around three sides of the obelisk are shaped like tree stumps painted rusty brown. Holes for ropes lay empty and flecked paint reveals the wooden texture of the forms. Long ago someone carefully carved tree rings into the bluntly cut tops. A few graves are scattered behind the obelisk—some tiny enough for babies. Tombstones rest behind stone-ringed graves; one is for Baby Jerry Ogata, his gravestone fresh and new, giving evidence that Manzanar still comforts broken hearts. All but one lay with heads pointing towards the Sierra Nevada and feet stretching down to the Owens River. The odd one lies alone facing south, with only a fence post for a tombstone. Plants recently placed on graves have dried and someone has pruned off the blossoms. The plastic plants hold their blooms though they, too, fade in the dry sun. The cemetery has been tended over the years and stands out in stark contrast to the crumbling camp ruins.

The ruins at Manzanar convey the incredible efforts made by Japanese Americans to go on about their lives despite confinement. Many saw their internment as a duty, their contribution to America's war efforts. Jack K. Semura volunteered to enter Manzanar early and use his carpentry skills to help build the barracks. He left his family, who followed later. Mrs. Semura recalled: "My mother and father always taught us honesty and justice and patriotism and all that to your country where you are born, so I figured, well, if they put me out, why can't I take a smile and take it bravely. That's the attitude I had. So, with a smile I left, and when I entered I said, 'So this is it. I have to make the best of it.'"[34] She lived three months with her infant son in a barracks with strangers before she was allowed to move into a barracks with her husband. Nevertheless, the camp had ornamental and vegetable gardens, orchards, and all the ingredients of a small rural town "complete with schools, churches, Boy Scouts, beauty parlors, neighborhood gossip, fire and police departments, glee clubs, softball leagues, Abbott and Costello movies, tennis courts, and traveling shows," even high school dances where girls leaned shyly against the wall waiting to be plucked onto the dance floor by the boys.[35] There was every accouterment of an American town, another Bedford Falls, except none of the town folk could leave, making for a not-so-wonderful life.

Perhaps this is what makes the internees' efforts seem noble and their ruins feel so poetic. They, like many Americans, tried to bring their landscape traditions into the desert. Surrounded by strangeness, they cultivated familiarity. Ansel Adams visited Manzanar in the fall of 1943 to photograph the relocation center for his book *Born Free and Equal* (1944). As someone who spent many years photographing the Sierra Nevada, someone who, in his own words, was "striving to reveal by the clear statement of the lens those qualities of the natural scene which claim the emotional and spiritual response of the people," Adams was uniquely poised to record the poetic connections between landscape and human experience at Manzanar:

> From the harsh soil they have extracted fine crops; they have made gardens glow in firebreaks and between the barracks. Out of the jostling, dusty confusion of the first bleak days in raw barracks they have modulated to a democratic internal society and a praiseworthy personal adjustment to conditions beyond their control. The huge vistas and stern realities of the sun and wind and space symbolize the immensity and opportunity of America—perhaps a vital reassurance following the experiences of enforced exodus.[36]

This reaction to the new, the urge to drape it with the old and familiar, almost always causes ecological problems in the West—primarily because water, the most precious element in the Western landscape, is drained from its natural cycle to irrigate these imported oases. Yet at Manzanar it feels justified and poignant, because the people had no other choice. They could not pack up and go home, so they tried to bring a flavor of their world to Manzanar. The gardens denied the region's landscape ecology, which allowed internees to forget for a moment where they were.

Many of the Japanese Americans interned in Manzanar owned land along the California coast and had produced successful crops. They carried their skills to the desert and harvested enough produce to support their confined city, the largest single settlement the Owens Valley has ever known. Kango Takamura remembers the transformation of Manzanar: "Just snakes, such a wild place! . . . so hot, you see, and the wind blows. There's no shade at all. It's miserable, really. But one year after, it's quite a change. A year after they built the camp and put water there, the green grows up. And mentally everyone is better."[37] Like the settlers of Manzanar who came before, internees insulated themselves from the desert with fields and gardens. They brought their land practices and land ethic with them into a strange place and did what they could to transform it into a home they could understand and recognize. Only Manzanar's golf links, with its sifted-soil putting greens, stood in surrender to the surrounding desert. Although probably of small concern for the internees, chipping onto these greens one must have had a hard time telling if the ball was on or in a sand trap.

Desert has reclaimed this land over the last fifty years. Cottonwoods and willows—some of the hardiest planted species—still hang on, particularly toward the mountains. Fruit trees bear signs of neglect. Saltbush dominates throughout, except where there are large patches of bare ground. Perhaps thousands of human feet compacted the soil beyond its ability to spring forth new life. The relocation camp is vanishing into the sandy desert soil. Adams predicted this decay:

> When all the occupants of Manzanar have resumed their places in the stream of American life, these flimsy buildings will vanish, the greens and flowers brought in to make life more understandable will wither, the old orchards will grow

163

older, remnants of paths, foundations and terracing will gradually blend into the stable texture of the desert. The stone shells of the gateways and the shaft of the cemetery monument will assume the dignity of desert ruins; the wind will move over the land and the snow fall upon it; the hot summer sun will nourish the gray sage and shimmer in the gullies.[38]

As slow, yet persistent, vegetation growth gradually returns Manzanar to the desert, plants begin to heal this wound in American history. It's reprehensible that American citizens and their Issei parents—those born on Japanese soil—were contained in camps during the war merely because they were of Japanese descent, so it's comforting to see Manzanar seep into the desert dust, a dignified ruin.

MARCH 26. Manzanar plants. In a spot along a wash, an old concrete foundation juts out, cracked and collapsed by the stream undercutting. An old plumbing pipe hangs under the concrete looking as if it will still pour murky, rusty water into the dry wash. Tumbleweeds clog the wash, suggesting a recent flood. I turn the soil with the end of my sneaker and see darkened, moist sand break through the dry surface. The tumbleweed's blond twigs and bone-white main branches contrast beautifully with the soft gray-green of the surrounding saltbush, whose dried terminal blossoms hover over them like a golden hazy halo. The weedy stream, a comical expression of aridity and desert desolation, has an Alice-in-Wonderland kind of appeal. Although the tumbleweed scene is the result of invasion, disruption of the natural vegetation cover and stream dynamics, it is beautiful. Perhaps if I learned more about tumbleweed ecology I'd see the ugliness in the landscape, as I have in the Manzanar ruins by learning that the crumbling foundations, roads, and gardens that have so captivated me are, in fact, remnants of a concentration camp.

A LAND BETWEEN

The United States government has stepped in to prevent Manzanar's natural decay into desert by shifting its stewardship to the National Park Service and creating the Manzanar National Historic Site. Manzanar is an important part of United States history, but I wonder in what way it should be remembered. How valuable would a restored Manzanar be; would it bring to life the travesty of over fifty years ago or would it be too tidy and safe, thus obliterating the meaning and power of the ruins, making it seem like just another movie set in the desert? Could it ever be restored without 10,000 Japanese Americans packing the tiny barracks, tending gardens along the firebreaks, singing in the choir, checking the job board for a way out, tilling the fields, hiding behind cardboard boxes to use the communal toilets in private, or giving birth in the hospital?

Why remember Manzanar? Claiming Manzanar as a significant historic site has brought up controversy about whose history should be told and how it should be interpreted. Some Numu tribal members argued against the commemoration of the Japanese Americans who spent only three and a half years at Manzanar, since the length of Numu suffering and imprisonment was far longer. Their resentment is understandable in light of the much more devastating confinement of their ancestors at Fort Tejon. They also contested any land development because "to develop an elaborate Japanese-American project means the desecration of the spiritual cultural heritage of the aborigines."[39] The World War II center may seem more immediate or more connected to modern American culture, but when the Manzanar War Relocation Center was established, Numu babies who were among the people herded to Fort Tejon like cattle and left to starve would have been in their eighties. 1863 was not all that long ago. America is as responsible for the travesty of Numu incarceration as it is for the Japanese confinement at Manzanar.

Resistance to using the Manzanar site to memorialize and interpret Japanese American internment also has come from war veterans. A visit to the local cemeteries makes it clear that the valley has been home to many patriots; grave sites mark the deaths of soldiers from every American war of the twentieth century. Some families and comrades of these soldiers, particularly those from World War II, resent that the government has chosen to memorialize Japanese encampment rather than the American soldiers who fought overseas to protect their country from the Japanese. Remember instead Pearl Harbor, they say, when twenty-one ships went down, 2,403 Americans were killed, 1,178 were wounded, and the nation went to war. The National Park Service is endeavoring to remember both; in Hawaii, above the sunken U.S.S. *Arizona*, where soldiers still lie in their watery graves, floats a memorial to the people lost in the Pearl Harbor attack, and now along U.S. 395, the long human history of Manzanar is remembered. Volunteers for the Eastern California Museum give walking tours around the camp and talk of the Numu, settlers, and internees.

People make a pilgrimage to Manzanar each spring to remember the experience of internment. Begun in 1969 by those who had been internees, it is meant to remind people that places such as Manzanar existed and, lest we forget history, could exist again. Karl Yoneda, former internee and author of *Ganbatte: Sixty-Year Struggle of a Kibei Worker* (1983) spoke at the thirteenth pilgrimage and reminded the attendants and those who watched the national television news coverage that "Manzanar is everywhere, whenever injustice raises its ugly head. It is the Indian reservations with close to one million Native Americans still contained in them; it is the ghettos where thousands upon thousands of racial minorities are shunted; it is the prisons where thousands are confined because most of them are poor and of different color and race."[40] The land has become symbolic of cultural and racial prejudice. It can be remembered and reflected upon because it is a ruin and, as a ruin, Manzanar helps American society face its own racism.

MARCH 12. Another visit to Manzanar. On this incredible morning, the air crisp and cold, it feels as if Manzanar remembers all the people who have walked its dusty reaches. The cemetery monument glistens as white as the Sierra peaks in the early morning sun. The breeze alternates between scents of sweet spring blossoms and cow dung. Birds are waking and a band of coyote give an eerie communal wail to the sun as it rises over the Inyo and light shifts across the valley floor. Frost patterns on old wood melt away, leaving damp patches to dry. Manzanar remembers with rusted tin cans and bread pans holding ice tilted in their half-sunken places, rusted spigots sticking up from the decomposed granite at haphazard angles, and concrete foundations with plumbing holes that are all that remain of the latrines where modest Japanese women went in 1942 to pee. Manzanar remembers with the garden resting in a ring of dead trees and scattered with cow droppings. "Aug 9, 1942" is written in the concrete at the edge of the tiny bridge. It was the dead heat of summer in the desert when internees poured the last bit of concrete for the water garden. The trees would be tiny then, but fifty years of their fallen leaves now turn to murky debris in the pools. Still water, calm as death from the last big rain, reflects the blue sky. A rabbit hops from rabbitbrush to rabbitbrush in the distance. Rusted and bent remains of a rail stick up on each corner of the bridge.

Manzanar remembers with the hospital and morgue's cold slab foundations all the babies born here, the families begun here, the last breaths taken here. Toward the cemetery, along tumbleweed-clogged washes, wild roses cling to last year's wilting hips while spring foliage breaks. In air filled with stillness and silence, the hum of distant planes and highway traffic is swallowed in the quiet. There is a magic here . . . as though ghosts linger like the rosehips and the ruins, recalling the care that was given to the land. Manzanar remembers with the remnant gardens growing around the hospital, where an old fruit tree still blooms white. Each year the gullies erode away the gardens, helping Manzanar to forget.

Manzanar speaks in wailing wind and calm silence. It speaks of regeneration—of slowly letting flood and wind till under this poor crop. The desert has infinite patience; it'll churn these ruins up and wash them down the aqueduct to Los Angeles—a gritty, yet just, dessert. Look west to see what rain, cold, heat, and wind can do to even the mightiest of mountains. Manzanar will be desert again.

168

170

There it is. Take it.

WILLIAM MULHOLLAND, SPEAKING TO A CROWD OF 30,000 AT THE

NOVEMBER 5, 1913 LOS ANGELES AQUEDUCT OPENING CEREMONY

"Army of Occupation" at the Alabama Gates, November 16, 1924
(*above*). Los Angeles Aqueduct, 1991 (*below*).

Ludwig Linde and his son Lawson beside an artesian well, before World War I.

Tumbleweed wash at Manzanar National Historic Site.

Mount Whitney Fish Hatchery.

Grandpa George Fish showing off a rainbow trout. Behind him,
great-grandpa Ben Fish—in a Stetson, as always, ca. 1950s.

I worked on Mule Train, *too, did I tell you that? Broke all them mules.*
They were all pack mules and we had to break 'em to harness.
PETE OLIVAS, IN DAVE HOLLAND'S *ON LOCATION IN LONE PINE*

Packer Jim Brumfield and a string of mules heading home, Sierra Nevada.

Song of the West movie set near the Alabama Hills, 1930.

Town of Manzanar, ca. 1925 (*above*). Manzanar War Relocation
Center barracks during occupation, ca. 1942–45 (*below*). Mount
Williamson in background.

Sometimes in the evenings we could walk down the raked gravel paths. You could face away from the barracks, look past a tiny rapids toward the darkening mountains, and for a while not be a prisoner at all. You could hang suspended in some odd, almost lovely land you could not escape from yet almost didn't want to leave.

JEANNE WAKASUKI HOUSTON AND JAMES D. HOUSTON, *FAREWELL TO MANZANAR*

Manzanar War Relocation Center, two little girls walking through Pleasure Park near blocks 33 and 34 during occupation, ca. 1942–45.

Manzanar National Historic Site, remnants of the hospital steps.

Manzanar National Historic Site, cemetery monument (*above*).
Offerings left on the cemetery monument at Manzanar (*below*).

EPILOGUE PARTING GLANCES

I T IS DIFFICULT to keep pace with history, so I've struggled to come to the end of this story of Owens Valley. With each visit to the valley I see changes—the widening of U.S. 395 between Lone Pine and Independence, a new billboard or warehouse on the outskirts of the small towns, a casino on reservation land. While the mountains, streams, and forests seem constant, the sense of how people interact with the land and one another is changing. "This is the nineties, nearly 2000, things are changing, becoming more cooperative," says a DWP public relations man. He's referring to how the Los Angeles and local agencies now interact, but his sentiment reflects more than this long and often contentious relationship. The place that once seemed to change only by means of conflict now is managed by a complex web of agency and citizen actions. For example, catching a fish at Lake Sabrina, up Bishop Creek Canyon, seems simple enough, but it actually involves several organizations: the area is overseen by the U.S. Forest Service; the dam that enlarged the lake is owned and operated by Southern California Edison; the fish are supplied by the California Fish and Game and raised at one of the local fish hatcheries; the fish in the creek are raised on a private ranch and stocked through a program funded by local merchants and citizens. Changes in land use occur by council, committee, collaboration, and cooperation. This network structure flies in the face of the romantic image of the rugged individual, the industrious entrepreneur, or the sovereign Native American. It seems unlikely any future conflict in the valley will rally the same kind of gun-toting, bomb-flinging rabble-rousers as in the days of the water wars. The human inclination to organize now drives most land-use changes in and around Owens Valley and has tamed the Wild West feeling of the place.

The biggest controversy now seems to be about where to put the plaque at the Manzanar National Historic Site that memorializes the place as the first of ten "concentration camps . . . bounded by barbed wire and guard towers, confining 10,000 persons. The majority being American citizens." It pleads that "the injustices and humiliation suffered [at Manzanar] as a result of hysteria, racism, and economic exploitation never emerge again."[1] Twenty-six years after its dedication, some want the plaque on display as a historical artifact when the visitor center is opened. Others, most of them offended by the term *concentration camp*, want the plaque relegated to a remote spot on the site or removed all together. But even this emotional issue is being played out through public hearings and environmental impact statement procedure. The land and, even more so, the people whose lives and livelihood are intertwined with the land are reaching a level of acceptance of the ways things are rather than fighting for what might be. Like a person finally coming to terms with her own mortality, the people of Owens Valley seem to be settling into a state of calm.

The land has a way of soothing the human heart. I see it in the way strangers smile when they look toward the mountains or gaze past their fishing line to a coyote on the other side of the creek. I see this in myself. Changes in my own life during the nearly ten years I've roamed the valley alter the way I read the land, the way

I perceive changes in the land. While writing this book, I followed love to Arizona, got a dog, bought a house, became a professor, wife, and mother. I began this book while living in Berkeley, California, a town that tries so hard to be liberal it has become conservative. When my father settled my sisters, brother, and me in an old brown shingled house on Berkeley's Ashby Avenue, men were landing on the moon and college students were rioting for the end of the Vietnam War. Although lunar landings and the war have long since ceased, and reasons for riots are less compelling (e.g., the building of a sand volleyball court in People's Park), the town has remained fundamentally the same. I walked under the same awnings, passing the same shops on my way to Mrs. Monroe's first-grade class as I did to attend graduate school. Berkeley's urban environment evolves only by complete consensus (or revolt) of its eclectic populace, so the earthquake fault upon which the community sits creeps faster than decisions pass through the town council. Berkelian philosophy dictates that any new human intervention on the land is probably a bad one, a belief predicated on a general lack of faith in people, especially organized people. I brought this ideology with me on my first visit to Owens Valley, so entrenched was I in Berkeley dogma. The first draft of this book was littered with this dim view of humanity and its impact on the natural environment.

Now, as I write the end to this storytelling, I live in Metropolitan Phoenix, Arizona—a place so conservative, it's liberal. At the rate people are flocking to Phoenix, the entire population of Owens Valley could relocate to the area in about four months. Since I've moved to Phoenix, over half a million people have followed behind me. Half the region's population has lived in the Valley of the Sun for less than five years. Earthmovers clear the land, destroying an acre of Sonoran Desert every hour of every day to make room for all these newcomers. The pace only abates for

PARTING GLANCES

supply and labor shortages; last year it was framers, today it's dry-wall. Change in the land *is* the constant for a place so open to development that a river can be turned into a lake. When I first moved to Phoenix, I cried almost every time I drove the six-lane arterials that form the massive urban grid, always getting lost for lack of anything on which to hang my mental map. Nothing sat still long enough for me to get a fix on it. Yet, as the years go by, I grow more accustomed to change.

Owens Valley, geographically between Berkeley and Phoenix, offers a balance of Berkeley's stasis and Phoenix's rapid change that has become emblematic of the New West. In this place between, this midway, I note additions and subtractions to the land with less skepticism than I once did. I see human interventions as a kind of dialog with the earth, a never-ending conversation that maintains the bond between people and the land. If I, like geographer Ruth Baugh, were to return to the valley twenty years hence, the snow-cloaked Sierra would still rise above the trees, the bristlecone pines would still cling to the rocky dolomite in the White Mountains. Such is the land's enduring nature. Yet between the mountains, shifts in the sandy soil and changes in human lives would be innumerable—children grown and gone off to college, old-timers passed on, cattle turned to beef. The capacity of the Owens Valley to seem simultaneously immutable and alterable is one of the many comforting ironies imbedded in the land.

The natural and cultural history of the Owens Valley is filled with irony, odd twists that have helped me shed my narrow and negative perception of the acculturation of nature. The valley's slim profile and steep relief seem to force contrary things together—Manzanar and Independence, cattle and backpackers, alpine fell-fields and desert sand, banal conversations and majestic views from Whitney Peak, meadow wildflowers and twisted pines, Numu subjugation and settlers' loss of land, abundant water and aridity, the City of Los Angeles and wilderness recreation. The Owens Valley is nothing if not a land of irony. The contrasts and contradictions that can be found half-buried in the salty bottom-lands, above the timberline in the Sierra, or on the dusty dry lakebed reveal the compelling bond between people and place, exposing the false separation between nature and culture and holding the key to preserving the land.

As I come to the end of my storytelling, I remember what had gripped me about Owens Valley in the first place, what had compelled me to return time and again. Remembering the beginning has made it easier to reach the end. Mary Austin credits the "rainbow hills, the tender bluish mists, the luminous radiance of the spring" for the valley's "lotus charm."[2] The surprise for me is that it is not just the land but the people and their remnant markings on the land that lures me back. This irony for someone like me, always more at ease with plants than people, also is a comfort. I feel more linked to people in the American West for having lingered so long in Owens Valley. I first went to the valley expecting to see a dismal and forlorn place, a land whose only human marks were scars left by greedy exploitation of earth's gifts. I saw these things—abandoned mines, dams and canals, fallow fields, desiccated river and lake—but I also caught glimpses of a human bond to the earth, a spiritual link. I heard it in the voices of people who spoke of the land before, saw it in the way it appears in mountain scenes that decorate cemetery headstones, touched it in the stones that edged rotting paths at Manzanar. I read it in the land, in local's faces, in words written about the valley. For the first time in my life, people and how they altered the land captivated me. I read Reyner Banham's *Scenes in America Deserta* (1989) with as much hunger as I had once devoured John Muir's misanthropic writing. Banham's question, "If this is a desert, what are all these *people* doing here?" became my guiding mantra.[3] This book has really been my attempt to answer this very question. In doing so, I've lost much of my despair about American society, about human destruction of the earth.

Recently, in Albuquerque, New Mexico, at a gathering of J. B. Jackson's devotees after his death, I listened to William Least Heat-Moon, author of *Blue Highways* (1982), speak about writing, travel, and the American landscape. He said "to preserve land . . . we must restore our spiritual link to the land. We can do this by linking to those who have gone before us. The past is where we can find out who we've been."[4] And, I would argue, who we can become. His words described ideas that had been festering in my own mind, that glancing into the past, reading history in the folds of the land, gives roots to the here and now and grants us hope for the future. And this reflection preserves land not by extracting people but by including and remembering them. He reminded me that looking back was a way of moving forward, a way of reestablishing the conversation with the earth. In coming to the end, I think about why I chose to write about the valley, why I wasn't content to just take in the splendid views and then go home. I write because I no longer feel alone in my faith in the human connection to the earth. If this bond meant nothing, if it was an allowable sacrifice for human progress, I would have been alone on the top of Mount Whitney, alone on the walk through Schulman Grove in the Ancient Bristlecone Pine Forest, alone at Manzanar, and alone at the river's edge. Instead, I am simply one of thousands who come from all over the world to visit the Owens Valley each year. The crowds give me hope.

Writing about Owens Valley has also reminded me why I became a student of landscape in the first place—to become a better teacher, to help children preserve their natural ability to speak to the earth, and to help adults remember their own dialogs with dirt. Writing in and about Owens Valley has been my conversation with the land, with its people, and with its past. The other day I got out of my car at my mother-in-law's home in Phoenix and was greeted by a hummingbird. It flitted about my face, full of threats and boasts. I stood still, watching, enjoying the gift of the bird's tiny company. Isabel asked, "You talking to hummingbirds, Mommy?" I laughed at her innocent assumption that I could talk to birds, as though it were as easy as saying hello to a friend. Isabel's ability to converse with plants, animals, rocks, and streams is what I covet most as a parent. It is the spirit of Rachel Carson's plea to nurture a "sense of wonder" in the earth.[5] Writing about Owens Valley has helped lift the angst I sometimes feel for American society and its disconnection to the land. Every time I return to the Owens Valley, to the "dwelling place of a Great Spirit," I am buoyed up with hope.[6] Walking amongst the Alabamas, riding a trail up the steep escarpment of the Sierra, sitting in a meadow thick with wildflowers, or standing near the 4,700-year-old Methuselah Tree, I can only feel contentment and joy. Any despair for humanity, for the state of the earth, is supplanted by immediate and intimate sensations. I feel again the simple wonder of childhood, the innocent satisfaction in talking to hummingbirds, old trees, and stone. This gift gives me solace during my most recent visit to the valley.

JULY 13 AND 14. Some surprises. Though the valley has become familiar to me, I see some places for the first time while camping up Bishop Creek Canyon, driving the dirt road through the Buttermilk Country near Bishop, and sitting among blooming larkspur, swamp onions, and corn lilies in the Onion Valley. Over the years, I have read or heard talk of each place, so my mental map of the land is well developed—completely flawed, but well developed. I meet each place with surprise.

Up Bishop Creek Canyon, beside the stream, on the lakes,

fishing is the thing to do. "Any luck?" a stranger asks, naturally assuming we are here to fish. Why make the steep climb up State Route 168, why stand beside the Bishop Creek waters, if not to fish? Strings of trout dangle from the fists of anglers as they walk to their cars, catching their fill by midmorning. Lake Sabrina sits in a cup, the largest and last of a string of glacially formed lakes that feed the middle fork of Bishop Creek. At 9,180 feet, in this east-facing canyon, scarlet and blue penstemon, sulfur flower, angelica, tiger lily, yarrow, iris, and Indian paintbrush still bloom near the water in midsummer. Enlarged by a hydroelectric dam around 1904, the lake stores the cascading creek waters that once powered the mining towns of Tonopah and Goldfield, Nevada, 110 miles across Owens Valley and over the White Mountains. The unassuming earth dam that holds back water in Lake Sabrina seems an unlikely beginning for boomtown lights. A drive through Goldfield makes it hard to imagine a time when such a huge effort would have been profitable. More people rest in the Goldfield cemetery than remain in this self-proclaimed living ghost town.

At Lake Sabrina I expected to find wall-to-wall urban anglers, the hoards wearing "BITE ME" tee shirts who swarm the bakery in Bishop, but instead the Tuesday morning crowd is a quiet handful. The lake is stocked with about a thousand pounds of fish each week. The big fish, the lunkers, are raised by Tim Alpers, a local trout rancher, and planted through the Adopt-A-Creek program. A mother and her grown daughter fish from the dam, cigarettes hung loosely from theirs mouths, faces tanned and leathery, hands swollen, fingernails trimmed short and dirty on their functional, well-worked hands. A little girl on her first fishing trip has caught three trout, her father tells me proudly. Dangling from a cord, they are far too novel to go untouched by my daughter, Isabel. I expect to see Bishop Creek Canyon, the angler's paradise touted in local fishing and tourist guides, teaming with vacationers trying to hook their dinner. "Fishing drives the economy," says Beth Porter of the Eastern California Museum, so I'm surprised to find space and tranquility in one of the fishing hot spots near Bishop.

The Buttermilk road turns off the 168 just below a powerhouse, another piece of the hydroelectric system that harnesses Bishop Creek energy. The tranquil name, a relic from the region's more active dairying days, is deceiving, and I am again surprised by the land around me. It's hard to picture dairy farmers serving buttermilk in the 1880s to people who passed through this rugged region. Like the urban fishermen I didn't find at Lake Sabrina, no Heidiesque milch cow landscape can be found in the Buttermilk Country. Instead dry sage and bitterbrush blanket the rocky terrain, and mounds of boulders like those in the Alabamas rise from the desert scrub. *Buttermilking* is the local term used by climbers to describe their scramblings on these blond stones. Some cattle still graze the meadows that dapple green here and there among the more abundant brush. The road rambles in a loop, ending in a heartstopping stretch along the ridge of a fingery moraine that divides Bishop and Birch Creek watersheds. A local raised his brow when I told him my little family had tackled the road in a two-wheel-drive vehicle. Having crossed the bridgeless McGee Creek, wound through the birch and conifer forest—branches scraping against both sides of the car, and made the steep descent off the moraine, I understand his surprise. A mountain biker decked in red Lycra, who passed us on the road, said his friend's vehicle had been stuck for six hours in a gorge somewhere in the Buttermilk Country. He continued on to Bishop to get help.

Another glacially carved canyon south of Bishop holds Independence Creek, the stream that runs through the town bearing the same name. Two falls carve straight white lines down the bowl that form the Onion Valley, where the streams feed into Independence Creek. The chilled air is full of the sound of falling water. I thought the valley would be one big round meadow, like those of the Kern

Plateau, a sea of wildflowers fit for a Julie Andrews solo, her arms out-stretched in the grassy expanse. Instead, elfin patches of larkspur, swamp onion, and corn lily lay surrounded by taller willow. Sage-brush crowds the high slopes where a trail winds up to Kearsarge Pass and over the eastern range to join the John Muir Trail. Hikers doing the JMT, so called by those who speak of it frequently, come over the pass to Onion Valley, hitch a ride to town to pick up a cache of freeze-dried food at the post office. The campground at Onion Valley is plastered with warning signs about bears, who, with the countless ground squirrels and marmots, augment their natural diet with camp grub.

JULY 15. At the Independence Cemetery. Even the older gravesites of this well-cared-for cemetery are marked with large mar-ble headstones and obelisks. One grave is framed like a farmhouse by a low green picket fence. Concrete slabs simply reading *unknown* mark the unidentified or forgotten dead. A man from England is hav-ing DNA tested to see if the body buried in one such plot is a de-ceased ancestor. Beyond the cemetery border, a new building blocks the otherwise consoling view of the Sierra and lower valley. Cloth flowers adorn the graves; a white bunch of daisies lay on the grav-elly mound of one of the unknowns. A dragonfly skims the surface of an artificial pond before flitting off beyond the cemetery. Sitting here, looking out at the new construction, listening to wind chimes above a nearby grave mingle with the downshifting eighteen-wheel-ers on Highway 395, I sense the majesty of the mountains, the magic of the waning light mixing together with all the human clutter—phone lines, irrigation pipes, chain-link fencing, barbed wire, cars, trucks, and roads. The valley landscape feels spiritually gripping, yet familiar and funky. This land would be alluring if all were granite and grandeur, but it has an enduring draw because the magnificent ter-rain is completely intertwined with human ordinariness, the natural splendor blended with the cultural commonplace. Mount Whitney belongs here as much as the gravestones that remember a loving husband and father and a beloved six-year-old son that now is *only sleeping.*

After midnight at the Winnedumah Inn in Independence. Isabel and Joe are asleep in the next room. The mosquitoes and cold wind convinced us to forgo camping in the Onion Valley (elevation 9,200 feet). On my first trip to Owens Valley, I stayed at the Winnedumah Inn, and in the quiet hours past midnight I think back on that first trip. I can hear whispers of memory, the windows rattling in a storm, the rustle of scratchy woolen blankets, the creaking wooden floor. Back then I had no idea of the long relationship I would have with the valley. I had come to study plants that grew beside the river but had sat up late, tucked under musty blankets to keep out the early spring cold, writing about picket fences, abandoned mines, and graveyards. The inn has had a face-lift since that first visit, my room now cheerfully decorated with patchwork, rose stencils, and white lace. The antique bed, table, and dresser are trimmed in green and pink. On my first visit I talked to no one, preferring to wander the dirt roads alone with my books, too shy to bother strangers. Today I knocked on someone's kitchen door to ask her about Numu words.

JULY 16. Leaving Owens Valley. Winding up Westgard Pass to cross into Nevada, I look back toward the valley like someone looks back to a friend before boarding a plane, as though this parting glance can guarantee the bond of friendship will sustain the separation. The valley floor seems dusted brown and I can see the glaciers high in the Sierra Range. Alfred Lord Tennyson wrote in Ulysses (1833) "I am a part of all that I have met." To have met the Owens Valley land, water, plants, and people is a gift. As a landscape historian and writer, I have come to the end of my storytelling and do so with assurance that the story of Owens Valley is without end. I know the valley can survive many more years of human tramplings and, by doing so, many more people can reestablish their own forgotten conversation with the land. I whisper a goodbye for now and the Owens Valley becomes a shrinking reflection in my rear-view mirror.

Before the sun, moon at zenith, going home

Pine shadows stretch across a meadow as the sun
crests a bald mountain, streaking pines,
dawning on dewy wildflowers. Boulders dropped
like marbles amongst trees, rattle in the chill.
Hot coffee, cold morning, going home.

Mules packed, horses saddled, Jim thinking
about whiskey, and snow on New Army pass.
Toes frozen in my boots, we cross the stream;
Coco doesn't stop to drink. She knows
our trail leads east toward home and oats.

Our mules behave like dogs, howling
when they're left alone. Creatures of habit,
wanting their food the same time every day.
Not dogs, no tails wag, no tongues loll
just hooves echo over granite, reaching
for the rising sun, going home.

Dust clouds in shafts of the long light sun
creeping down the canyon. Willows, onions,
swamp whiteheads, yarrow, groundsel, purple aster
crowd the water's edge, resting in early
morning warmth that breaks frost.
Hot coffee, cold morning, going home.

MAPS OF OWENS VALLEY

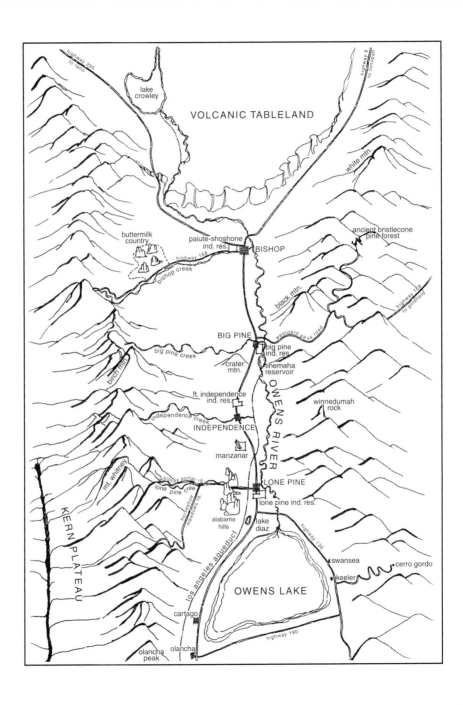

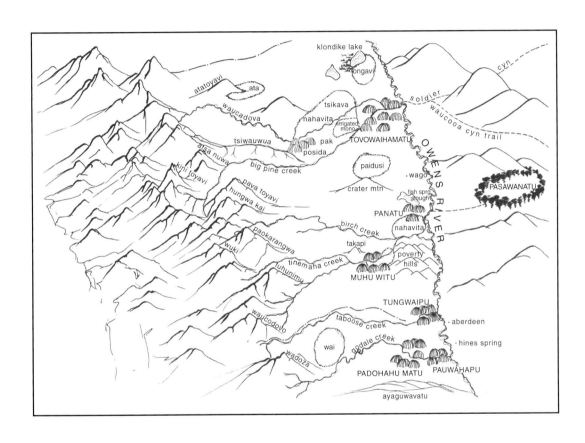

Owens Valley (*opposite page*) and Numu (Paiute) Village Territories around Big Pine (*above*). Map of Numu village territories adapted from maps in Julian Steward, "Ethnography of the Owens Valley Paiute," *American Archaeology and Ethnology* 33, no. 3 (1933): 233–350.

NOTES

INTRODUCTION

1. Julian H. Steward, "Myths of the Owens Valley Paiute," *University of California Publications in American Archaeology and Ethnology* 34, no. 5 (1936): 355–440; quote from 364. Steward's footnotes indicate that Hoavadunuki (Jack Stewart) is the source. Direct quotes and some source titles use other terms, such as *Indians*, *Owens Valley Paiute*, *savages*, or *redskins*, to describe the native people of Owens Valley. Since they called themselves *Numu* before any of these other words were applied to them, this is the name I use throughout the book.

The Numu language is traditionally not written. Before the nineteenth-century disruption of the native settlements, each village in the region had its own variations in pronunciation and definition of words. Ethnographers who have studied the Numu have tended to spell words phonetically and according to a unique, rather than a universal, system. All ethnographic research I found had been conducted after the Numu were relegated to reservation land. These factors have produced numerous inconsistencies in spelling between and even within sources. Throughout the book, I have used the most common spelling. For words published with diacritics, I have followed Steward's pronunciation guides to develop a simplified spelling. Julian H. Steward, "Ethnography of the Owens Valley Paiute," *American Archaeology and Ethnology* 33, no. 3 (1933): 233–350.

2. Myth printed in Willard D. Johnson, "A Legend of Owens Valley and a Geological Parallel," *Sierra Magazine* (Jan. 1909). This periodical was a bimonthly

publication entitled *Inyo Magazine* through December 1908 and was titled the monthly *Sierra Magazine* thereafter.

3. Zenas Leonard, *Narrative of the Adventures of Zenas Leonard* (1934; reprint, Lincoln: University of Nebraska, 1978), 196, 199, and 202; excerpts from Leonard's journal, 1831–35.

4. Ibid., 205.

5. Letter from Talbot to his mother, sent from Monterey, Alta California, July 24, 1846 in Robert V. Hine and Savoie Lottinville, eds., *Soldier in the West: Letters of Theodore Talbot During His Services in California, Mexico, and Oregon, 1845–53* (Norman: University of Oklahoma Press, 1972), 37.

6. Appendix Q, "Journal of Mr. Edward M. Kern of an Exploration of Mary's or Humboldt River, Carson Lake, and Owens River and Lake, in 1845" in *Report of Explorations Across the Great Basin of the Territory of Utah for a Direct Wagon-Route from Camp Floyd to Genoa, in Carson Valley, in 1859* by Captain J. H. Simpson (1876; reprint, Reno: University of Nevada Press, 1983), 482.

7. Ibid., four excerpts, 483, 482, and 484, respectively.

8. Letter of orders to Captain Davidson from Lieutenant Colonel Benjamin L. Beall, July 19, 1859, in Philip J. Wilke and Harry W. Lawton, eds., *The Expedition of Captain Davidson from Fort Tejon to Owens Valley in 1858* (Socorro, N. Mex.: Ballena Press, 1976), 14. I have included misspelled words as they appeared but omitted the editors' use of *sic*.

9. Ibid., 24–25. Five excerpts from Davidson's "Report of the Results of an Expedition to Owen's Lake, and River, with the Topographical Features of the Country, the Climate, Soil, Timber, Water, and also, the Habits, Arms, and Means of Subsistence, of the Indian Tribes seen upon the March. July and August, 1859."

1 A LAND BETWEEN

1. John Muir, *The Mountains of California* (1894; reprint, New York: Penquin Books, 1985), 39.

2. Clarence King, *Mountaineering in the Sierra Nevada* (1872; reprint, Philadelphia: J. B. Lippincott, 1963), three excerpts, 277 and 279.

3. Ibid., 279.

4. Muir, "A Rival of the Yosemite: The Cañon of the South Fork of King's River California," *Century Illustrated Monthly Magazine* 43 (Nov. 1891): 77–97; two excerpts, 93.

5. Anna Mills Johnston, "A Trip to Mt. Whitney in 1878," *Mt. Whitney Club Journal* 1, no. 1 (1902): 18–28, quoted in Leonard Daughenbaugh, "On Top of Her World: Anna Mills Ascent of Mt. Whitney," in *Mountains to Desert: Selected Inyo*

Readings (Independence, Calif.: Friends of the Eastern California Museum, 1988), three excerpts, 90 and 91.

6. Henry McLauren Brown, *Mineral King Country: Visalia to Mount Whitney* (Fresno, Calif.: Pioneer Publishing Co., 1988), 43. Excerpt from Edward Parsons's report on the 1903 Sierra Club outing, n.p., n.d.

7. Muir, "A Rival of the Yosemite," 93, 95.

8. Brown, *Mineral King Country,* 23.

9. Mary Austin, *The Land of Little Rain* (1903; reprint, Garden City, New York: Doubleday and Co., 1962), 1. Austin is best known for this book, which describes the land and people of the Owens Valley region. She lived in the valley towns of Lone Pine, Independence, and Bishop from 1892 to 1905.

10. Translation found in Mary DeDecker, *Mines of the Eastern Sierra* (Glendale, Calif.: La Siesta Press, 1993), 42.

11. Austin, *Land of Little Rain,* 2.

12. W. Newton Price, "The Circuit Rider," in Southern Inyo American Association of Retired People, *Saga of Inyo County* (Covina, Calif.: Taylor Publishing Co., 1966), 81.

13. John Glanville Dixon, "Brothers and Brides 'Go West,'" ibid., 121.

14. The first of a ten-part series entitled "A Theft in Water," *Inyo Magazine,* Sept.–Dec. 1908, pts. 1–8; in the renamed periodical *Sierra Magazine,* Jan.–Feb. 1909, pts. 9–10.

15. "The Story of the Irrigated Farm," *Sierra Magazine,* Jan. 1909.

16. Wallace Stegner, *The American West as Living Space* (Ann Arbor: University of Michigan Press, 1987), 8.

17. T. E. Jones, "Owens Lake in 1885," in SIAARP, *Saga,* 10.

18. Letter from Brigadier General George Wright to Colonel Ferris Forman, May 2, 1862, and letter from Lieutenant Colonel George S. Evans to Major Richard C. Drum, July 9, 1862, in *The War of the Rebellion: A Compilation of the Official Records of the Union and Confederate Armies,* ser. 1, vol. 50, pt. 1: "Reports, Correspondence, etc." (Washington, D.C.: Government Printing Office, 1897), 1047 and 148, respectively.

19. Muir, *The Mountains of California,* 234.

20. Excerpt from Brewer's journal in Francis P. Farquhar, ed., *Up and Down California in 1860–1864: The Journal of William H. Brewer, Professor of Agriculture in the Sheffield Scientific School from 1864 to 1903* (Berkeley: University of California Press, 1966), 535.

21. Elizabeth Carrasco, as told to Bernice Etcharren, "The Aqueduct," in SIAARP, *Saga,* 65.

2 DWELLING BEFORE

1. Wallace Stegner, *The American West as Living Space* (Ann Arbor: University of Michigan Press, 1987), 25.

2. Julian H. Steward, "Ethnography of the Owens Valley Paiute," *American Archaeology and Ethnology* 33, no. 3 (1933): 233–350; quote from 237.

3. Charles W. Campbell, "Origins and Ethnography of Prehistoric Man in Owens Valley," First paper of the *Eastern California Museum Anthropological Papers* (Independence, Calif.: March 1974), 6.

4. Stegner, *American West,* 24.

5. Steward, "Ethnography," 235.

6. Ibid., 308.

7. Ibid., 325–28; see the legends for two maps of Owens Valley, which mark numerous places significant to the Numu. For information on pronunciation and spelling, see Steward's "Ethnography" or consult with the Paiute Shoshone Indian Cultural Center in Bishop, California.

8. Steward, "Two Paiute Autobiographies," *American Archaeology and Ethnology* 33, no. 5 (1934): 423–38; three excerpts from Hoavadunuki's autobiography, 426 and 428.

9. Ibid., 430.

10. Ibid., 433 and 435; two excerpts from Sam Newland's autobiography.

11. News correspondent referred to only as "Quis," in Philip J. Wilke and Harry W. Lawton, eds., *The Expedition of Captain Davidson from Fort Tejon to Owens Valley in 1858* (Socorro, N. Mex.: Ballena Press, 1976), 33. The article originally appeared in the *Los Angeles Star,* Aug. 27, 1859.

12. Davidson's report in Wilke and Lawton, *The Expedition,* 26 and 27.

13. Ibid., 30.

14. Zenas Leonard, *Narrative of the Adventures of Zenas Leonard* (1934; reprint, Lincoln: University of Nebraska, 1978), 197, 200, and 199, respectively.

15. Letter from Lieutenant Colonel George S. Evans to Major Richard C. Drum, April 29, 1862, in *The War of the Rebellion: A Compilation of the Official Records of the Union and Confederate Armies,* ser. 1, vol. 50, pt. 1: "Reports, Correspondence, etc." (Washington, D.C.: Government Printing Office, 1897), 49.

16. Personal conversation with Mike Patterson, co-owner of Cerro Gordo.

17. Robert C. Likes and Glenn R. Day, *From This Mountain—Cerro Gordo* (Bishop, Calif.: Chalfant Press, 1975), 53.

18. Mary Austin, *Earth Horizon: Autobiography* (New York: Literary Guild, 1932), 233.

19. Steward, "Ethnography," 297. This phrase is set in quotes by Steward,

which implies it is the words verbatim of one of his informants. He doesn't note who is the speaker.

20. Report by Lieutenant Colonel George S. Evans, July 1, 1862, in *The War of the Rebellion,* pt. 1, 146. Though Evans uses no quotes in the letter, he implies these are the words of the Numu. "White man" is used to describe all non-Numu people who were settling ("sitting down") in the valley. This emigrant population was ethnically diverse and included women, but "white man" was the common catch-all phrase of the day.

21. Letter from Evans to Drum, April 29, 1862, ibid., pt. 1, 49.

22. Report by Evans, July 1, 1862, ibid., pt. 1, 146.

23. Letter from Brigadier General George Wright to Colonel Ferris Forman, May 2, 1862, ibid., pt. 1, 1047.

24. Letter from Lieutenant Colonel George S. Evans to Major Richard C. Drum, July 9, 1862, ibid., pt. 1, two excerpts, 148 and 149, respectively.

25. Report by Evans, July 1, 1862, ibid., two excerpts, 146.

26. Letters from Lieutenant Colonel George S. Evans to Major Richard C. Drum (Drum was promoted to Lieutenant Colonel by the time of the second letter), June 6, Oct. 7, and Sep. 16, 1862, ibid., pt. 1, 1121, 153, and 149, respectively.

27. Letter from Colonel James H. Carleton to Colonel George W. Bowie, March 17, 1862, in *The War of the Rebellion,* pt. 1, two excerpts, 936.

28. An assayer named Hanks (no first name given), quoted in Willie A. Chalfant, *The Story of Inyo* (1922, 1933; Bishop, Calif.: Chalfant Press, 1975), 178 and 183, respectively.

29. Letter from Major John M. O'Neill to Major Richard C. Drum, Aug. 8, 1862, in *The War of the Rebellion,* pt. 2, 75.

30. Mary Harry, "Biography of Mary Harry," in Frederick Seymour Hulse, *Big Pine Paiute Ethnographic Notes: Books 15–17;* and Mary Rooker, "Autobiography of Mary Rooker," in Hulse, *Fort Independence Paiute Ethnographic Notes: Book 18* (Bancroft Library, University of California at Berkeley, 1935), microfilm CU-23.1: 2216, items 92.3 and 93, frames 41 and 63, respectively. Hulse doesn't include his interviewees' Numu names. Mary Rooker identifies herself as Siea Wiu Neaka.

31. Harry, "Biography," CU-23.1:2216, item 92.3, two excerpts, frame 43.

32. Jennie Cashbaugh, "Biography of an Old Woman," in Hulse, *Bishop Paiute Ethnographic Notes: Books 7–14,* microfilm CU-23.1: 2216, item 91, Cashbaugh speaking, frames 374 and 375.

33. Letter from Captain John C. Schmidt at Fort Tejon to Colonel Richard C. Drum, Jan. 26, 1864, in *The War of the Rebellion,* pt. 2, 733.

34. Letter from Lieutenant Colonel Richard C. Drum to Brigadier General George Wright, Dec. 7, 1864, ibid., pt.2, 1085. W. Chalfant notes in *Story of Inyo*

that Joaquin Jim occupied the Southern Mono region during the conflicts and that he may have been originally from the west side of the Sierra.

35. "The Indian at Home," *Sierra Magazine,* Jan. 1909, n.p.

36. Cashbaugh, "Biography," frame 377.

37. "The Indian at Home," n.p.

38. Chalfant, *The Story of Inyo,* 229.

39. Helen S. Giffen, "Camp Independence: An Owens Valley Outpost," *Historical Society of Southern California* 23–24 (1941–42): 128-142, excerpt from 137.

40. A. W. Von Schmidt, field notes, July 15, 1855, quoted in Chalfant, *The Story of Inyo,* 121.

41. "The Call of the Soil," and "A New Magazine in a New Field," *Inyo Magazine,* Sept. 1 and Dec. 15, 1908, respectively.

42. Genny Schumacher Smith, ed., *Deepest Valley: A Guide to Owens Valley Its Roadsides and Mountain Trails* (Los Altos, Calif.: William Kaufman, 1962; rev. eds. 1978 and 1995), 191 (page citation to 1978 edition).

43. Harold P. Simonson, ed., introduction to Frederick Jackson Turner, *The Significance of the Frontier in American History* (New York: Continuum Publishing Co., 1990), 9. Original paper presented in 1893.

44. C. Mulholland, "The Owens Valley Earthquake of 1872," *Annual Publication of the Historical Society of Southern California* 5 (1894): 27-31; two excerpts, 27 and 29, respectively.

45. Excerpt from Circular No. 8, War Department, Surgeon General's Office, May 1, 1875, in Giffen, "Camp Independence," 139.

46. Gussie M. Wood, "The Kispert Ranch on Georges Creek," in Southern Inyo American Association of Retired People, *Saga of Inyo County* (Covina, Calif.: Taylor Publishing Co., 1966), 113.

47. Davidson's report in Wilke and Lawton, *The Expedition,* 27.

48. "The Call of the Soil," n.p.

49. Turner, *The Significance of the Frontier,* 39.

50. John E. Jones, "A Pioneer Record," in SIAARP, *Saga,* two excerpts, 116.

51. Clara Shaw Eddy, "Harry Shaw, 1869–1925," ibid., 117.

52. Robert A. Sauder, *The Lost Frontier: Water Diversion in the Growth and Destruction of Owens Valley Agriculture* (Tucson: University of Arizona Press, 1994), 63. The following figures for agricultural production are also from Sauder. He relied on the agricultural census for this information.

53. "Inyo, a Virgin Empire," *Inyo Magazine,* Aug. 15, 1908.

1. Mary Austin, *The Land of Little Rain* (1903; reprint, Garden City, New York: Doubleday and Co., 1962), 53.

2. Both phrases were popularized in the era of boosterism (during the late nineteenth to the early twentieth century) and are indicative of the religious undertones of booster ideology. Captain Davidson notes that the Numu have made the desert "blossom as the rose" in his report in Philip J. Wilke and Harry W. Lawton, eds., *The Expedition of Captain Davidson from Fort Tejon to Owens Valley in 1858* (Socorro, N. Mex.: Ballena Press, 1976), 20. The Owens Valley is referred to as the "land of milk and honey" in "Inyo, a Virgin Empire," *Inyo Magazine,* Aug. 15, 1908.

3. Mary Austin, *Earth Horizon: Autobiography* (New York: Literary Guild, 1932), 308.

4. Ruth E. Baugh, "Land Use Changes in the Bishop Area of Owens Valley, California," *Economic Geography* 13 (Jan. 1937): 17– 34; excerpt from 26.

5. Ibid., 17.

6. Art Hess, who came to the valley in 1907, quoted in Los Angeles Department of Water and Power, "The Owens Valley Controversy in Perspective," in Southern Inyo American Association of Retired People, *Saga of Inyo County* (Covina, Calif.: Taylor Publishing Co., 1966), 43. Original quote appeared in the *Los Angeles Times,* June 20, 1973, in an article by sportswriter Lupi Saldana.

7. This phrase was the foundation of early conservation practices and is attributed to Gifford Pinchot, founder of the U. S. Forest Service.

8. "The Los Angeles Water Deal," *Inyo Magazine,* Sep. 1, 1908, the foreword to a ten-part series entitled "A Theft in Water" (1908–9) that chronicled the events leading up to diversion.

9. Ibid.

10. William Mulholland, quoted in Abraham Hoffman, *Vision or Villany: Origins of the Owens Valley–Los Angeles Water Controversy* (College Station: Texas A & M University Press, 1981; reprint 1992), 172.

11. Excerpt from a letter from President Theodore Roosevelt to Secretary of the Interior Hitchcock, June 25, 1906, in "The Los Angeles Water Deal," *Inyo Magazine,* Sep. 1, 1908.

12. "A Theft in Water, 5," *Inyo Magazine,* Nov. 1, 1908.

13. Beveridge Ross Spear, "Water Everywhere," and Vera T. Jones, "Manzanar," both in SIAARP, *Saga,* 50 and 106, respectively.

14. Beveridge Ross Spear, "Owens Valley—in the Beginning," and W. Newton Price, "The Circuit Rider," in SIAARP, *Saga,* 9 and 83, respectively.

15. Genny Schumacher Smith, ed., *Deepest Valley: A Guide to Owens Valley Its Roadsides and Mountain Trails* (Los Altos, Calif.: William Kaufman, 1962; rev. eds. 1978 and 1995), 197.

16. Ibid., 198, and Anonymous, "The Dow Villa," in SIAARP, *Saga,* 26.

17. Henry McLauren Brown, *Mineral King Country: Visalia to Mount Whitney* (Fresno, Calif.: Pioneer Publishing Co., 1988), 13.

18. The Olivas cabin at Monache Meadow was scheduled for demolition in 1996 but is now being preserved through the cooperative efforts of the Backcountry Horsemen of California and the National Forest Service.

19. Ethel and Henry Olivas, as told to B. C. Dawson, "The Olivas Family," in SIAARP, *Saga,* two excerpts from 57. Context indicates the words are those of Henry Olivas.

20. Clarence King, *Mountaineering in the Sierra Nevada* (1872; reprint, Philadelphia: J. B. Lippincott, 1963), 246.

21. Quoted in Brown, *Mineral King Country,* 18; most likely from a personal conversation between Brown and John Crowley.

22. Charley Morgan speaking in "Oral Interview: Charley Morgan and Henry Brown," conducted by Loiuse Jackson Snyder, Oct. 9, 1993, Springville, California, transcript, 6. My uncle, Henry Brown, sent me the transcript without noting where the original recording was housed.

23. Leslie A. Hancock, "My Dog Was a Movie Actor," in SIAARP, *Saga,* 53.

24. Bill Turque, "The War for the West," *Newsweek Magazine,* Sep. 30, 1991, 18–35; two quotes, 24.

25. Martha L. Mills, "Henry Lenbek Family—Manzanar," in SIAARP, *Saga,* 123.

26. Michi Weglyn, *Years of Infamy: The Untold Story of America's Concentration Camps* (New York: Morrow Quill Paperbacks, 1976), 53, quoting Dr. Eugene V. Rostow, and 38, respectively.

27. Ritsuko Eder oral history, interview conducted by Catherine Piercy, May 21, 1973 (Independence, Calif.: Eastern California Museum), transcript OH009.1, 5.

28. Harlan D. Unrau, *The Evacuation and Relocation of Persons of Japanese Ancestry during World War II: A Historical Study of the Manzanar War Relocation Center,* vol. 1 (Denver, Colo.: National Parks Service, NPS D-3, 1996), 99.

29. The Eastern California Museum in Independence has an extensive collection of artifacts from the Manzanar War Relocation Center, including a chair made from scrap lumber.

30. George Fukasawa in Arthur A. Hansen, ed., *Japanese American World War II Evacuation Oral History Project,* pt. 2: "Internees" (Westport, Conn.: Meckler Publishing, 1991), 236.

31. Yoriyuki Kikuchi quoted ibid., 206.

32. Kango Takamura quoted in Deborah Gesensway and Mindy Roseman, *Beyond Words: Images from America's Concentration Camps* (Ithaca, N.Y.: Cornell University Press, 1987), 123.

33. Versions of this poem appear in Gesensway and Roseman, *Beyond Words,* and Sue Kunitomi Embrey, *The Lost Years, 1942–1946* (Los Angeles: Moonlight Publications, 1972; reprint 1987), 108 and 35, respectively (Embrey page citation from the reprint edition). The version that appears here is from the original, held in the Cornell Library archives.

34. Mrs. Jack K. Semura (only formal name noted) oral history, interview conducted by Catherine Piercy, Aug. 17, 1973 (Independence, Calif.: Eastern California Museum), transcript OH34.1, 6.

35. Jeanne Wakasuki Houston and James D. Houston, *Farewell to Manzanar* (New York: Bantam Books, 1973), 72.

36. Ansel Adams, *Born Free and Equal* (New York: U.S. Camera, 1944), 9.

37. Takamura quoted in Gesensway and Roseman, *Beyond Words,* 123.

38. Adams, *Born Free and Equal,* 25.

39. Letter from Ruth Brown, Jessie Durant, Florence Espinueva, Walter Parsons, and Blanche Shippentower, Owens Valley Tribal Elders, to the Inyo County Board of Supervisors, May 22, 1979, 1. Photocopy in author's collection.

40. Karl G. Yoneda, *Ganbatte: Sixty-Year Struggle of a Kibei Worker* (Los Angeles: University of California Asian American Studies Center Resource Development and Publications, 1983), 201.

EPILOGUE   PARTING GLANCES

1. The plaque was placed at the Manzanar National Historic Site by the State Department of Parks and Recreation in cooperation with the Manzanar Committee and the Japanese American Citizens League, April 14,1973. It designates Manzanar as California Registered Historic Landmark number 850.

2. Mary Austin, *The Land of Little Rain* (1903; reprint, Garden City, New York: Doubleday and Co., 1962), 6.

3. P. Reyner Banham, *Scenes in America Deserta* (1982; reprint, Cambridge: MIT Press, 1989), 158.

4. William Least Heat-Moon, in a lecture given at the J. B. Jackson and American Landscape Conference, Albuquerque, New Mexico, Oct. 1–4, 1998.

5. Rachel Carson, *The Sense of Wonder* (New York: Harper and Row, 1956).

6. Translation found in Mary DeDecker, *Mines of the Eastern Sierra* (Glendale, Calif.: La Siesta Press, 1993), 42.

Full citations are given in the notes for all references from which I quote directly. The following are selected sources I found particularly informative while researching the natural and cultural history of the Owens Valley. I begin with sources that are general in content and then suggest some references relevant to specific topics. I also include movies filmed in the valley; they lend insight into how the landscape is perceived and, in some cases, are just plain entertaining.

OWENS VALLEY GENERAL INTEREST

The most useful book on the Owens Valley landscape and history is *Deepest Valley*, edited by Jeff Putnam and Genny Schumacher Smith (1962, 1978; Mammoth Lakes, Calif.: Genny Smith Books, 1995). I carried an earlier edition (1978) on all my roamings, and when the new edition came out I felt less guilty about ripping the old edition up so I could carry the plant section to Mount Whitney summit without bringing along excess weight. Other sources for general history and geography for the valley include the Inyo County Board of Supervisors' brief book *Inyo 1866–1960* (Bishop, Calif.: Chalfant Press, 1966), Willie A. Chalfant's *The Story of Inyo* (Bishop, Calif.: Chalfant Press, 1922), *Mountains to Desert—Selected Inyo Readings* (Independence, Calif., 1988), compiled and published by the Friends of the Eastern California Museum, Sue Irwin's *California's Eastern Sierra: A Visitor's Guide* (Los Olivos, Calif.: Cachuma Press, 1991, rev. ed., 1997), and Helen Hoffman's *Owens Valley: A Guide to Independence and Lone Pine* (Bishop,

Calif.: Chalfant Press, 1984). Mary Austin's precious little book *The Land of Little Rain* (1903; Garden City, N.Y.: Doubleday and Co., 1962) contains classic essays on life in the Inyo County region and poetic descriptions of natural and human activities, phenomenon, and peculiarities of the Owens Valley and surrounding territory. It's still in print; the Penguin edition (New York, 1988) has an introduction by Edward Abbey and is also available on cassette, read by Terry Tempest Williams, which is a treat for the car-traveler exploring the many dirt roads in the valley and mountains. In Independence, the Eastern California Museum is also a great storehouse for history of the Owens Valley. It has a library of regional references, many oral histories of old-timers and Manzanar internees in the valley, and an extensive photograph collection.

MOUNTAINS

Much of the natural history of the Sierra Nevada I learned growing up, but I relied on a few sources while out in the field or writing at home. My first book lessons on the mountains came from Francis P. Farquhar's *History of the Sierra Nevada* (Berkeley: University of California Press, 1969) and John Muir's *The Mountains of California* (1894; New York: Penguin Books, 1985). Both give a sense of the mountains as they were first experienced by early American emigrants. Muir's love of the alpine terrain is clear in his series of poetic essays. Having read his piece on the water ouzel, I was thrilled to recognize the unusual bird flitting through the river-spray along the Kern. My uncle, Henry McLauren Brown, taught me many of the wonders of the Southern Sierra. His books, *Mineral King Country: Visalia to Mount Whitney* (Fresno, Calif.: Pioneer Publishing Co., 1988) and *Kern Canyon Country* (Porterville, Calif.: self-published, 1991), present the history of one of his most loved regions with the same personal familiarity he put into my lessons along the Kern, when he taught me about glacial moraines, old mountain men, and soda springs and had me hug the biggest ponderosa pine in the canyon.

Clarence King's *Mountaineering in the Sierra Nevada* (1872; Lincoln: University of Nebraska Press, 1997) conveys the nineteenth-century perception of these Western mountains; his descriptions are amply packed with enough sublimity and grandeur to match an Albert Bierstadt painting. I carried *Sierra Nevada Natural History* by Tracy I. Storer and Robert L. Usinger (Berkeley: University of California Press, 1963) in my saddlebag, pulling it out constantly along the trails that traverse the Kern Plateau. Two little references helped me in hiking to the top of Mount Whitney: Walt Wheelock and Tom Condon's *Climbing Mount Whitney*, 5th ed. (Glendale, Calif.: La Siesta Press, 1995) and Thomas Winnett's *High Sierra*

Hiking Guide: Mt. Whitney, 2d ed. (Berkeley: Wilderness Press, 1992). Ralph Cutter's *Sierra Trout Guide* (Portland: Frank Amato Publications, 1991) contains lovely renderings of trout by Joseph R. Tomellen and rich information on trout ecology and behavior. *Natural History of the White-Inyo Range, Eastern California*, edited by Clarence A. Hall, Jr. (Berkeley: University of California Press, 1991) is a valuable source for information on plants in the Inyo Mountains. It also contains a great discussion by Robert L. Bettinger on Numu hunting and gathering practices. *The Ancient Bristlecone Pine Forest* by Russ and Anne Johnson (Bishop, Calif.: Chalfant Press, 1970; rev. ed., 1978) discusses the natural history of these old trees that grow in the White Mountains. For a survey of California's natural communities, Elna Bakker's *An Island Called California* (Berkeley: University of California Press, 1971; rev. ed. 1984) is helpful. I relied on *Basin and Range* by John McPhee (New York: Noonday Press, 1981), *The Field Guide of Geology* by David Lambert et al. (New York: Facts on File, Inc., 1988), and *Physical Geography of the Global Environment* by H. J. de Blij and Peter O. Muller (2d ed., New York: John Wiley, 1996) to answer general questions on the natural history of rocks, mountains, and valleys.

WATER

Water in the Owens Valley is a much-published topic, so I only list essential sources here. The most entertaining story told of the Los Angeles aqueduct is in *Cadillac Desert: the American West and Its Disappearing Water* by Marc Reisner (New York: Viking, 1986). A revised and updated edition was put out by Penguin (New York, 1993). By far the most thoroughly researched account of the valley's water wars with Los Angeles is given in William L. Kahrl's *Water and Power: The Conflict over Los Angeles' Water Supply in the Owens Valley* (Berkeley: University of California Press, 1982). Donald Worster also tackles the issue soundly in *Rivers of Empire: Water, Aridity and the Growth of the American West* (New York: Pantheon, 1985), as does John Walton in *Western Times and Water Wars: State, Culture, and Rebellion in California* (Berkeley: University of California, 1992). A balanced version of the story is Abraham Hoffman's *Vision or Villainy: Origins of the Owens Valley–Los Angeles Water Controversy* (College Station: Texas A & M University Press, 1981). Robert Sauder focuses on the effects of water diversion on agriculture in *The Lost Frontier* (Tucson: University of Arizona Press, 1994). To be fair, one might read *Sharing the Vision: the Story of the Los Angeles Aqueduct* (Los Angeles, 1988), put out by the Los Angeles Department of Water and Power. Blake Gumprecht's *The Los Angeles River: Its Life, Death, and Possible Rebirth* (Baltimore: Johns Hopkins University Press, 1999) offers an excellent glimpse into why Los

211

Angeles took water from the Owens Valley. Mary Austin offers a fictional version of the story, in which water stays in the valley, in *The Ford* (1917; Berkeley: University of California Press, 1997).

THE DESERT LANDSCAPE

For a discussion of the unpeopled deserts of the Owens Valley region, Edmund C. Jaeger's *The North American Deserts* (1957; Stanford, Calif.: Stanford University Press, 1981) and *The California Deserts* (1933; Stanford, Calif.: Stanford University Press, 1987) provide thorough information. For a peopled perspective, Reyner Banham's *Scenes in America Deserta* (Cambridge: MIT Press, 1989) gives a rich and intriguing view of the American deserts as seen by an intellectual tourist with an eye for art and architecture. He finds what Edward Abbey calls "the unity of opposites" in a barren landscape riddled with human artifacts. Patricia Nelson Limerick in *Desert Passages* (Niwot, Colo.: University Press of Colorado, 1989) also gives a human perspective of the desert through an analysis of American desert literature. Mary Austin's *Land of Little Rain* does as fine a job depicting people who lived and roamed the valley in the late nineteenth century as it does describing the natural history of the region's flora and fauna. *Daggett: Life in a Mojave Frontier Town*, edited by Peter Wild (Baltimore: Johns Hopkins University Press, 1997) presents a vivid account of human settlement in the arid reaches of California.

THE WEST AND CALIFORNIA

Owens Valley's place in the American West is obvious when reading Frederick Jackson Turner's *The Significance of the Frontier in American History* (1893; New York: Continuum Publishing Co., 1990). From his pre-reclamation-era perspective, he treats the desert more as a barrier than as a place to settle. Wallace Stegner, in *The American West as Living Space* (Ann Arbor: University of Michigan Press, 1987), updates Turner's frontier thesis in his excellent compact essays on aridity, water, and land use in the American West. Gerald D. Nash presents an analysis of changing historiographic perceptions of the West in *Creating the West: Historical Interpretations 1890–1990* (Albuquerque: University of New Mexico Press, 1991). John Brinckerhoff Jackson has been a constant inspiration in my studies of the American West, particularly *The Necessity for Ruins and Other Topics* (Amherst: University of Massachusetts Press, 1980) and *A Sense of Place, a Sense of Time* (New Haven, Conn.: Yale University Press, 1994). There are volumes upon volumes published on California history; I used John W. Caughey's *California:*

History of a Remarkable State, 4th ed. (Englewood Cliffs, N.J.: Prentice-Hall, 1982) for basic information. George R. Stewart's *The California Trail: An Epic with Many Heros* (Lincoln: University of Nebraska Press, 1962) and Dale L. Morgan's *The Humboldt: Highroad of the West* (1943; Lincoln: University of Nebraska Press, 1985) chronicle the developments along the overland trail to California. John C. Frémont fans might enjoy his bulky *Memoirs of My Life* (Chicago: Belford, Clarke, & Co., 1887), especially the 655-page first volume, which includes recollections of his five explorations in the West. For perspectives of overland emigrant women in the West, I looked to Dee Brown's *The Gentle Tamers: Women of the Old Wild West* (1958; Lincoln: University of Nebraska Press, 1985) and my own great-great-grandmother Juliette G. Fish's unpublished typescript, *Crossing the Plains in 1862* (1924). A copy of the Fish recollections is kept at the Carpinteria Historical Museum in Carpinteria, California. *Let the Cowboy Ride* by Paul Starrs (Baltimore: Johns Hopkins University Press, 1998) offers an excellent picture of cattle ranching in the West.

213

VALLEY DWELLERS

Julian H. Steward's ethnographic work with the Numu is invaluable, particularly *Ethnography of the Owens Valley Paiute* (Berkeley: University of California Press, 1933), in which he discusses food, hunting, gathering, irrigating, clothing, government, ritual, and social practices, and *Two Paiute Autobiographies* (Berkeley: University of California Press, 1934), which records two male Numu life stories. In Bishop, exhibits in the Paiute Shoshone Indian Cultural Center display traditional structures used by the Numu and describe their pinenut harvesting methods. While not an easily accessible source, the microfilm reels of Frederick Seymour Hulse's "Ethnographic Notes on Owens Valley Paiute" (Berkeley, Calif., Bancroft Library, 1935) contain many transcribed oral histories, including Numu recollections of the exodus to Fort Tejon during the Cival War. For the military perspective of the same era, *The War of the Rebellion: A Compilation of the Official Records of the Union and Confederate Armies,* ser. 1, vol. 50, parts 1 and 2 (Washington, D.C.: Government Printing Office, 1897) includes letters from soldiers who were involved in the subjugation of the California native peoples. The men included both regular army officers and soldiers from the California Volunteer battalions. Dorothy Clora Cragen's *The Boys in the Sky-Blue Pants: The Men and Events at Camp Independence and Forts of Eastern California, Nevada and Utah, 1862–1877* (Fresno, Calif.: Pioneer Publishing Co., 1975) gives an account of the military presence in the Owens Valley. The book is well documented from the perspective of the army and newly emigrated residents. As the title suggests, Cragen's version is a bit laundered; the Hulse oral histories and compiled army letters convey a grubbier, less sky-blue image of the conflict.

Accounts of early American forays into the valley that are most insightful are Zenas Leonard's *Adventures of Zenas Leonard* (1839; Lincoln: University of Nebraska Press, 1978), *The Expedition of Captain Davidson from Fort Tejon to Owens Valley in 1858*, edited by Philip J. Wilke and Harry W. Lawton (Socorro, N. Mex.: Ballena Press, 1976), and *Up and Down California in 1860–1864: The Journal of William H. Brewer, Professor of Agriculture in the Sheffield Scientific School from 1864 to 1903*, edited by Francis P. Farquhar (Berkeley: University of California Press, 1966). Two small but useful publications on mining are Mary DeDecker's *Mines of the Eastern Sierra* (Glendale, Calif.: La Siesta Press, 1993), and Robert C. Likes and Glenn R. Day's *From This Mountain—Cerro Gordo* (Bishop, Calif.: Chalfant Press, 1975). *A History of Hispanics in Southern Nevada* by Malvin Lane Miranda (Reno: University of Nevada Press, 1997) and *Aztlan: the Southwest and Its People* by Luis F. Hernandez (Rochelle Park, N.J.: Hayden Book Co., 1975) describe the Hispanic presence in the region's nineteenth-century mining industry. Remi Nadeau, a descendent of the man who hauled ore from Cerro Gordo to Los Angeles, relates the valley's mining developments to the growth of Los Angeles in *City-Makers* (Garden City, N.Y.: Doubleday and Co., 1948). Two sources that convey the settler's frame of mind and relationship to the land are *Inyo Magazine* (Independence, Calif.), which changed its name to *Sierra Magazine* in 1909, a periodical filled with boosterish descriptions of the valley's natural resources, and the Southern Inyo American Association of Retired People's *Saga of Inyo County* (Covina, Calif.: Taylor Publishing Co., 1966), a compilation of stories told by old-timers in the valley. *Inyo 1866–1966* (Bishop, Calif.: Chalfant Press, 1966), put out by the Inyo County Board of Supervisors, has many great historical photographs, particularly of the pioneer era.

MANZANAR

For a personal view of Manzanar, Jeanne Wakasuki Houston and James D. Houston's *Farewell to Manzanar* (New York: Bantam, 1974) tells the story through memories of Wakasuki Houston, who spent her childhood in the camp. Ansel Adams presents exquisite photographs of Manzanar during World War II in *Born Free and Equal* (New York: U.S. Camera, 1944). His photographs can also be seen in John Armor and Peter Wright's *Manzanar* (New York: Times Books, 1989), a small book documenting Manzanar's history. *Camp and Community: Manzanar and the Owens Valley*, edited by Jessie A. Garrett and Ronald C. Larson (Fullerton, Calif.: California State University, Oral History Program, 1977), a book of oral histories, gives a perspective of the camp from people who lived in the valley. Other oral histories can be found in *Japanese American World War II Evacuation Oral History Project*, edited by Arthur A. Hansen (Westport, Conn.: Meckler, 1991), and *Voices Long Silent: An Oral History into the Japanese American Evacuation*, edited by Hansen and Betsy E. Mitson (Fullerton, Calif.: California State University, Oral History Program, 1974). A compact history of the camp that includes a chronology for the relocation period is *The Lost Years: 1942–46*, edited by Sue Kunitomi Embrey (Los Angeles: Moonlight Publications, 1972). Harlan D. Unrau's *The Evacuation and Relocation of Persons of Japanese Ancestry During World War II: A Historical Study of the Manzanar War Relocation Center* (Denver, Colo.: National Parks Service, 1996) is a comprehensive two-volume history of the land, from Numu use through Japanese encampment. The *General Management Plan and Environmental Impact Statement: Manzanar National Historic Site* (San Francisco: National Park Service, 1996) considers options for developing visitor facilities. Deborah Gesensway and Mindy Roseman's *Beyond Words: Images from America's Concentration Camps* (Ithaca, N.Y.: Cornell University Press, 1987) has wonderful prints of artwork done by people while interned at Manzanar and the other camps.

FILMS

Films in which the Owens Valley landscape plays significant roles include *Gunga Din* (1939), the Hollywood version of Rudyard Kipling's verse; *Tycoon* (1947), which has John Wayne racing against nature's wrath to build a rail bridge in the Andes; *Springfield Rifle* (1952), a confusing story about Civil War counterespionage starring the ever-perplexed-looking Gary Cooper; and *How the West Was Won* (1962), John Ford's epic about westward migration in which a wagon train is chased by native peoples through the Alabamas. These four films afford great images of the Alabama Hills as India, the Andes, and the American West. For scenes of the bottomlands and more barren reaches of the valley around Owens Lake, see *Three Godfathers* (1948) starring John Wayne, perhaps the inspiration for the French film *Trois Hommes et un Couffin* (1985) and its American remake *Three Men and a Baby* (1987); *Nevada Smith* (1966), a hot dry western starring Steve McQueen; and *Star Trek V* (1989), a hot dry sci-fi adventure starring William Shatner. The mountains and their narrow winding roads are memorable in *High Sierra* (1941), with Humphrey Bogart's "Mad Dog" Earle heading for the hills in a high-speed chase up the Whitney Portal Road, and *The Long, Long Trailer* (1954), in which Lucille Ball and Desi Arnaz perform their usual antics up the same steep grade. Jack Palance portrays "Mad Dog" in the *High Sierra* remake, *I Died a Thousand Times* (1955). Owens Valley plays a foreign role as the Crimea in *Charge of the Light Brigade* (1936), starring Errol Flynn and Olivia De Havilland, and again as Kipling's India in *Kim* (1951), with Errol Flynn. The valley depicts a close approximation of itself in the sci-fi thriller *Tremors* (1989), with giant people-eating sandworms, and *Trial and Error* (1997), in which a lawyer finds true love with a small-town waitress.

For someone with a taste for Westerns, the film fare is abundant: *The Round Up* (1920), starring Fatty Arbuckle; *Boots n' Saddles* (1937), starring Gene Autry; *Broken Arrow* (1950), starring Jimmy Stewart; *Cow Town* (1950), again with Gene Autry; *Dynamite Pass* (1950), with Tim Holt; *Rawhide* (1951), featuring Susan Hayward and Tyrone Power; *Along the Great Divide* (1951), starring Walter Brennan and Kirk Douglas; *Hangman's Knot* (1952), with Donna Reed and Randolph Scott; *Valley of Fire* (1952), starring Gene Autry; *The Tall T* (1957), with Maureen O'Sullivan and Randolph Scott; *The Law and Jake Wade* (1958), with Robert Taylor; *North to Alaska* (1960), starring the Duke; and the more recent *Maverick* (1993), featuring Jodie Foster, James Garner, and Mel Gibson. The landscape has been featured on television in *Bonanza, Have Gun, Will Travel, The Lone Ranger, The Rockford Files,* and *The Virginian,* and it has increasingly become the landscape of choice for automobile commercials. A good source of the valley's film history is Dave Holland's *On Location in Lone Pine* (Granada Hills, Calif.: Holland House, 1990).

When I began writing this book, I was filling out keypunch cards to check out books from the library and using a Mac Plus, both arcane and abandoned writing tools in this new century of electronic information. Nowadays, anyone interested in trolling for information about the valley, especially regarding municipal, agency, or association activities (e.g., the town of Bishop, the National Forest Service, or the Backcountry Horseman's Association), can go online. Internet sources are useful for gathering basic information, but I opted not to list web sites relevant to the Owens Valley because, better than any electronic information, the printed and screen sources helped me sense the landscape, hear the valley's many voices, feel the deep history that rests with the layered alluvium on the bottomlands. Another scholar of the valley might name more sources, especially on the water issues, but the ones listed here helped me hear the valley's heartbeat. That is enough.

218

INDEX

221

REBECCA FISH EWAN was born in Santa Barbara. Her ancestors immigrated to Sandwich, Massachusetts, in 1637 and gradually moved westward, finally coming overland to California in 1862. For four generations, the Fish family grew lima beans for their seed company near Santa Barbara. Rebecca moved to Berkeley in 1968. She received a bachelor's degree in mathematics and a master of landscape architecture from the University of California at Berkeley. She was awarded a grant from the Graham Foundation for Advanced Studies in Fine Arts for *A Land Between*. She is an assistant professor of landscape architecture at Arizona State University and lives in Tempe with her husband, Joe, and daughter, Isabel. This is her first book.